GEOL

a Golden Guide® from St. Martin's Press

by
FRANK H. T. RHODES

Illustrated by
RAYMOND PERLMAN

St. Martin's Press ♒ New York

FOREWORD

"THE EARTH—love it or leave it," is a popular slogan evolving from our recognition of the perils of an exploding population, a polluted environment, and the limited natural resources of an already plundered planet. Effective solutions to our societal problems demand an effective knowledge of the earth on which we live.

The object of this book is to introduce the earth: its relation to the rest of the universe, the rocks and minerals of which it is made up, the forces that shape it, and the 5 billion years of history that have given it its present form. Many topics that are discussed in the book have a very practical and urgent significance for our society, things such as water supply, industrial minerals, ore deposits, and fuels. Others are of importance in longer planning and development, such as earthquake effects, marine erosion, landslides, arid regions, and so on. The earth provides the ultimate basis of our present society: the air we breathe, the water we drink, the food we eat, the materials we use. All are the products of our planet.

The study of the earth opens our eyes to a vast scale of time that provides us with new dimensions, meanings, and perspectives. And it reveals the order of the earth, its dynamic interdependence and its structured beauty.

I am most grateful to Sharon Sanford, who typed the manuscript of the revised edition.

F.H.T.R.

ISBN 1-58238-143-7

CONTENTS

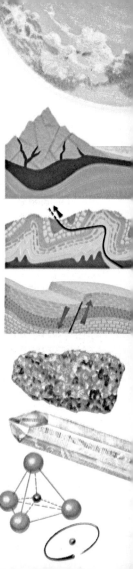

Volcanic eruption in Kapoho, Hawaii. Many islands in the Pacific are volcanic in origin.

GEOLOGY AND OURSELVES

Geology is the study of the earth. As a science, it is a newcomer in comparison with, say, astronomy. Whereas geology is only about 200 years old, astronomy was actively studied by the Egyptians as long as 4,000 years ago. Yet speculation about the earth and its activities must be as old as the human race. Surely, primitive people were familiar with such natural disasters as earthquakes and volcanic eruptions.

Gradually, human society became more dependent upon the earth in increasingly complex ways. Today, behind the insulation of our modern living conditions, civilization remains basically dependent upon our knowledge of the earth. All our minerals come from the earth's crust. Water supply, agriculture, and land use also depend upon sound geologic information.

Geology stimulates the mind. It makes use of almost all other sciences and gives much to them in return. It is the basis of modern society.

THE BRANCH OF GEOLOGY emphasized here is physical geology. Other Golden Guides of this series, *Rocks and Minerals* and *Fossils*, deal with the branches of mineralogy, petrology, and paleontology.

PHYSICAL GEOLOGY is the overall study of the earth, embracing most other branches of geology but stressing the dynamic and structural aspects. It includes a study of landscape development, the earth's interior, the nature of mountains, and the composition of rocks and minerals.

HISTORICAL GEOLOGY is the study of the history of earth and its inhabitants. It traces ancient geographies and the evolution of life.

ECONOMIC GEOLOGY is geology applied to the search for and exploitation of mineral resources such as metallic ores, fuels, and water.

STRUCTURAL GEOLOGY (tectonics) is the study of earth structures and their relationship to the forces that produce the structures.

GEOPHYSICS is the study of the earth's physical properties. It includes the study of earthquakes (seismology) and methods of mineral and oil exploration.

PHYSICAL OCEANOGRAPHY is closely related to geology and is concerned with the seas, major ocean basins, seafloors, and the crust beneath them.

THE SIZE AND SHAPE OF THE EARTH were not always calculated accurately. Most ancient peoples thought the earth was flat, but there are many simple proofs that the earth is a sphere. For instance, as a ship approaches from over the horizon, masts or funnels are visible. As the ship comes closer, more of its lower parts come into view. Final proof, of course, was provided by circumnavigating the globe and by photographs taken from spacecraft.

The Greek geographer and astronomer Eratosthenes was probably the first (about 225 B.C.) to measure successfully the circumference of the earth. The basis for his calculations was the measurement of the elevation of the sun from two different points on the globe. Two simultaneous observations were made, one from Alexandria, Egypt (Point B, p. 7), and the other from a site on the Nile near the present Aswan Dam (Point A). At the latter point, a good vertical sighting could be made, as the sun was known to shine directly down a well at noon on the longest day (June 23) of the year.

Eratosthenes reasoned that if the earth were round, the noonday sun could not appear in the same position in the sky as seen by two widely separated observers. He compared the angular displacement of the sun (Y) with the distance between the two ground sites, A and B.

RECENT DATA from orbiting earth satellites have confirmed that the earth is actually slightly flattened at the poles. It is an oblate spheroid, the polar circumference being 27 miles less than at the equator. The following measurements are currently accepted:

Avg. diameter	7,918 mi.
Avg. radius	3,959 mi.
Avg. circumference	24,900 mi.

LARGE AS THE EARTH IS, it is minute in comparison with the universe, where distances are measured in light years—the distance light, moving at 180,000 miles per second, travels in a year. This is about 6 billion miles or 10 million, million kilometers. Using these units of measurements, the moon is 1.25 light *seconds* from the earth, the sun is 8 light *minutes* from the earth, and the nearest star is 4 light *years* from the earth. Our galaxy is 80,000 light years in diameter. The most distant galaxies are 8 billion light years from earth. It is estimated that there are at least 400 million galaxies "visible" from earth using radio telescopes and similar means of detection. Galaxies are either elliptical or spiral in shape.

ERATOSTHENES measured the distance (X) between Points A and B as 5,000 stadia (about 575 miles). Although the observer at Point A saw the sun directly overhead at noon, the observer at B found the sun was inclined at an angle of 7° 12' (Y) to the vertical. Since a reading of 7° 12' corresponds to one-fiftieth of a full circle (360°), Eratosthenes reasoned that the measured ground distance of 5,000 stadia must represent one-fiftieth of the earth's circumference. He calculated the entire circumference to be about 28,750 miles.

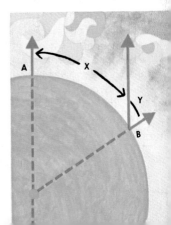

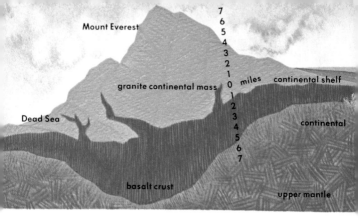

Mount Everest · granite continental mass · miles · continental shelf · Dead Sea · continental · basalt crust · upper mantle

THE EARTH'S SURFACE, for the purpose of measurements, is commonly assumed to be uniform, for mountains, valleys, and ocean deeps, great as some are, are relatively insignificant features in comparison with the diameter of the earth.

But mountains are not insignificant to humans. They play a major role in controlling the climate of the continents; they have profoundly influenced the patterns of human migration and settlement.

Mountain ranges, with very few exceptions, are narrow, arcuate belts, thousands of miles in length, generally developed on the margins of the ancient cores or shields of the continents. They consist of great thicknesses of sedimentary and volcanic rocks, many of them of marine origin. Their intense folding and faulting are evidence of enormous compressive forces.

Mountains are not limited to the land. The ocean floor has even more relief than the continents. Most of the continental margins extend as continental shelves to a depth of about 600 feet below the level of the sea, beyond

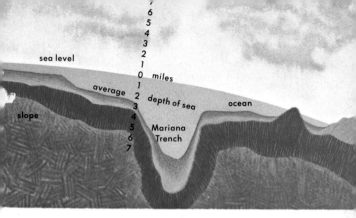

which the seafloor (continental slope) plunges abruptly down (see p. 66).

The ocean floor adjacent to some islands and continents has long, deep trenches, the deepest of which, off the Philippines, is about 6½ miles deep (p. 138). A worldwide network of mid-oceanic ridges, of mountainous proportions, encircles the earth. This network has geophysical and geologic characteristics that suggest it occupies a unique role in earth dynamics—that along these ridges new seafloor is constantly being created.

THE EARTH'S CRUST, derived from the denser, underlying mantle (pp. 128-129), consists of two kinds of rock. The continental crust differs from the oceanic crust in being lighter (2.7 gm./cc. compared with 3.0), thicker (35 km-70 km versus 6 km), older (up to 3.5 billion years versus a maximum of 200 million years), chemically different, and much more complex in structure. These differences reflect the different modes of formation of the two kinds of crust (pp. 140-145).

THE EARTH

EARTH is one of nine planets revolving in nearly circular (elliptical) orbits around our star, the sun. Earth is the third planet out from the sun and the fifth largest planet in our solar

PLANETS vary in size, composition, and orbit. Mercury, with a diameter of 3,112 miles, is the planet nearest to the sun. It orbits the sun in just three earth months. Jupiter, about ten times the diameter of Earth (88,000 miles), is the largest planet and fifth in distance from the sun, taking about 11¾ earth years to orbit the sun. Pluto, the most distant planet, takes about 247¾ earth years to orbit around the sun. The inner planets have densities, and probably compositions similar to Earth's; outer planets are gaseous, liquid, or frozen hydrogen and other gases.

THE SUN, an average-sized star, makes up about 99 percent of the mass of the solar system. Its size may be illustrated by visualizing it as a marble. At this scale, the earth would be the size of a small grain of sand one yard away. Pluto would be a rather smaller grain 40 yards away.

SATELLITES revolve around seven planets. Including the earth's moon, there are 61 satellites altogether; Mars has 2, Jupiter 16, Saturn 18, Uranus 15, Neptune 8, and Pluto 1.

COMETS are among the oldest members of the solar system. They orbit the sun in extremely long, elliptical orbits. As comets approach the sun, their tails begin to glow from friction with the solar wind.

RELATIVE DISTANCES OF PLANETS FROM THE SUN

3,000 million miles

2,000 million miles

1,000 million miles

Pluto

Neptune

Uranus

Saturn

Jupiter

Mars

Earth   Venus

Mercury

IN SPACE

Pluto

Mercury

Mars

Venus

Earth

system, having a diameter of about 7,918 miles. It completes one orbit around the sun in about 365¼ days, the length of time that gives us our unit of time called a year.

THE MOON, earth's natural satellite, has about ¼ the diameter, ¹⁄₈₁ the weight, and ³⁄₅ the density of our planet. The moon completes one orbit around the earth every 27⅓ days. It takes about the same length of time, 29½ days, to rotate on its own axis; hence, the same side, with an 18% variation, always faces us. The moon's surface, cratered by meteorite impact, consists of dark areas (maria) which are separated by lighter mountainous areas (terrae). Terrae are part of the original crust, formed about 4.5 billion years ago; maria are basins, excavated by meteorite falls, filled by basaltic lavas formed from 3 to 4 billon years ago.

Neptune

Uranus

ASTEROIDS, the so-called minor planets, are rocky, airless, barren, irregularly shaped objects that range from less than a mile to about 480 miles in diameter. Most of the asteroids that have been charted travel in elongated orbits between Mars and Jupiter. The great width of this zone suggests that the asteroids may be remnants of a disintegrated planet formerly having occupied this space.

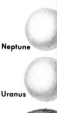

METEORS, loosely called shooting stars, are smaller than asteroids, some being the size of grains of dust. Millions daily race into the earth's atmosphere, where friction heats them to incandescence. Most meteors disintegrate to dust, but fragments of larger meteors sometimes reach the earth's surface as "meteorites." About 30 elements, closely matching those of the earth, have been identified in meteorites.

Saturn

Jupiter

11

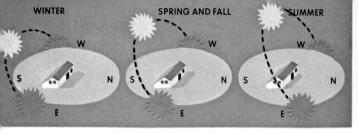

Tilted axis determines different positions of sun at sunrise, noon, and sunset at different seasons in middle north latitude.

THE EARTH'S MOTIONS determine the daily phenomenon of day and night and the yearly phenomenon of seasonal changes. The earth revolves around the sun in a slightly elliptical orbit and also rotates on its own axis. Since the earth's axis is tilted about 23½° with respect to the plane of the orbit, each hemisphere receives more light and heat from the sun during one half of the year than during the other half. The season in which a hemisphere is most directly tilted toward the sun is summer. Where the tilt is away from the sun, the season is winter.

RELATIVE MOTIONS OF THE EARTH

REVOLUTION is earth's motion about the sun in a 600-million-mile orbit, as it completes one orbit about every 365¼ days traveling at 66,000 mph.

ROTATION is a whirling motion of the earth on its own axis once in about every 24 hours at a speed of about 1,000 mph at the equator.

NUTATION is a daily circular motion at each of the earth's poles about 40 ft. in diameter.

PRECESSION is a motion at the poles describing one complete circle every 26,000 years due to axis tilt, caused by gravitational action of the sun and moon.

OUR SOLAR SYSTEM revolves around the center of our Milky Way Galaxy. Our portion of the Milky Way makes one revolution each 200 million years.

GALAXIES seem to be receding from the earth at speeds proportional to their distances.

THE SUN is the source of almost all energy on earth. Solar heat creates most wind and also causes evaporation from the oceans and other bodies of water, resulting in precipitation. Rain fills rivers and reservoirs, and makes hydroelectric power possible. Coal and petroleum are fossil remains of plants and animals that, when living, required sunlight. In one hour the earth receives solar energy equivalent to the energy contained in more than 20 billion tons of coal, and this is only half of one billionth of the sun's total radiation.

Just a star of average size, the sun is yet so vast that it could contain over a million earths. Its diameter, 864,000 miles, is over 100 times that of the earth. It is a gaseous mass with such high temperatures (11,000° F at the surface, perhaps 325,000,000°F at the center) that the gases are incandescent. As a huge nuclear furnace, the sun converts hydrogen to helium, simultaneously changing four million tons of matter into energy each second.

Solar prominences compared with the size of the earth

Earth

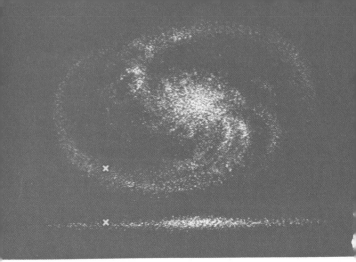

THE MILKY WAY, like many other galaxies, is a whirling spiral with a central lens-shaped disc that stretches into spiral arms. Most of its 100 billion stars are located in the disc. The Milky Way's diameter is about 80,000 light years; its thickness, about 6,500 light years. (A light year is the distance light travels in one year at a velocity of 186,000 mi. per sec., or a total of about 6 trillion miles.)

GALAXIES are huge concentrations of stars. Within the universe, there are innumerable galaxies, many resembling our own Milky Way. Sometimes called extragalactic nebulae or island universes, these star systems are mostly visible only by telescope. Only the great spiral nebula Andromeda and the two irregular nebulae known as the Magellanic Clouds can be seen with the naked eye. Telescopic inspection reveals galaxies at the furthermost limits of the observable universe. All of these gigantic spiral systems seem to be of comparable size and rotating rapidly. Nearly 50 percent appear to be isolated in space, but many galaxies belong to multiple systems containing two or more extragalactic nebulae. Our galaxy is a member of the Local Group, which contains about a dozen other galaxies. Some are elliptical in shape, others irregular. Galaxies may contain up to hundreds of billions of stars and have diameters of up to 160,000 light years. Galaxies are separated from one another by great spaces, usually of about 3 million light years.

Many galaxies rotate on their own axes, but all galaxies move bodily through space at speeds of up to 100 miles a second. In addition to this, the whole universe seems to be expanding, moving away from us at great speeds. Our nearest galaxy, in Andromeda, is 2.2 million light years away.

About 100 million galaxies are known, each containing many billions of stars. Others undoubtedly lie beyond the reach of our telescopes. It seems very probable that many of the stars the galaxies contain have planetary systems similar to our own. It has been estimated that there may be as many as 10^{19} of these. Chances of life occuring on other planets would, therefore, seem very high, although it may not bear an exact resemblance to life on earth.

WHIRLPOOL NEBULA in Canes Venatici, showing the relatively close packing of stars in the central part

GREAT SPIRAL NEBULA M31 in Andromeda is similar in form but twice the size of our own galaxy, the Milky Way.

THE CHEMICAL ELEMENTS are the simplest components of the universe and cannot be broken down by chemical means. Ninety-two occur naturally on earth, 70 in the sun. They develop from thermonuclear fusion within the stars, in which the elementary particles of the lightest elements (hydrogen and helium) are transformed into heavier elements.

THE ORIGIN OF THE UNIVERSE is unknown, but all the bodies in the universe seem to be retreating from a common point, their speeds becoming greater as they get farther away. This gave rise to the expanding-universe theory, which holds that all matter was once concentrated in a very small area. Only neutrons could exist in such a compact core. According to this theory, at some moment in time — at least 5 billion years ago — expansion began, the chemical elements were formed, and turbulent cells of hot gases probably originated. The latter separated into galaxies, within which other turbulent clouds formed, and these ultimately condensed to give stars. Proponents refer to this as the "Big Bang" theory, a term descriptive of the initial event, perhaps as long as 10-15 billion years ago.

THE ORIGIN OF THE SOLAR SYSTEM is not fully understood, but the similar ages of its components (Moon, meteorites, Earth at about 4.5-4.6 billion years) and the similar orbits, rotation, and direction of movement around the sun, all suggest a single origin. The theory currently most popular suggests that it formed from a cloud of cold gas, ice, and a little dust, which began slowly to rotate and contract. Continuing rotation and contraction of this disc-shaped cloud led to condensation and thermonuclear fusion — perhaps triggered by a nearby supernova, from which stars such as the sun were formed. Collision of scattered materials in the disc gradually led to the formation of bodies—planetismals—which became protoplanets. The growing heat of the sun probably evaporated off the light elements from the inner planets (now represented by the dense, rocky "terrestrial" planets—Mercury, Venus, Earth, Mars, and the Moon). The outer planets, because of their greater distance from the sun, were less affected and retained their lighter hydrogen, helium, and water composition. Perhaps they formed from mini-solar-planet systems within the larger disc. This composition may well reflect that of the parent gas cloud.

Each planet seems to have had a distinct "geologic" history. Some, like Earth and Io, a moon of Jupiter, are still active. Others, like Mercury, Mars, and our moon, had an earlier active history, but are now "dead."

This theory, in an earlier version, has a long history, going back to Immanuel Kant, the philosopher, in 1755, and the French mathematician Pierre-Simon de Laplace (1796).

Artist's interpretation of the dust-cloud theory

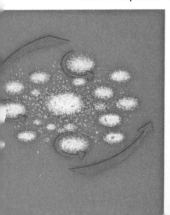

THE EARTH'S ATMOSPHERE is a gaseous envelope surrounding the earth to a height of 500 miles and is held in place by the earth's gravity. Denser gases lie within three miles of the earth's surface. Here the atmosphere provides the gases essential to life: oxygen, carbon dioxide, water vapor, and nitrogen.

Differences in atmospheric moisture, temperature, and pressure combined with the earth's rotation and geographic features produce varying movements of the atmosphere across the face of the planet, and conditions we experience as weather. Climatic conditions (rain, ice, wind, etc.) are important in rock weathering; the atmosphere also influences chemical weathering.

Gases in the atmosphere act not only as a gigantic insulator for the earth by filtering out most of the ultraviolet and cosmic radiation but also burn up millions of meteors before they reach the earth.

The atmosphere insulates the earth against large temperature changes and makes long-distance radio communications possible by reflecting radio waves from the earth. It also probably reflects much interstellar "noise" into space, which would make radio and television as we know them impossible.

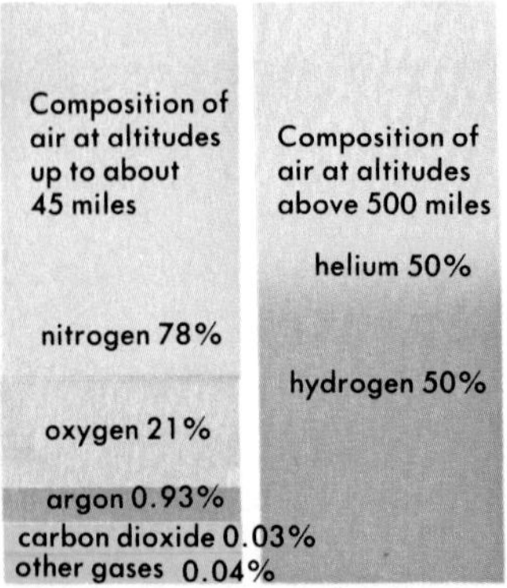

Composition of
air at altitudes
up to about
45 miles

nitrogen 78%

oxygen 21%

argon 0.93%
carbon dioxide 0.03%
other gases 0.04%

Composition of
air at altitudes
above 500 miles

helium 50%

hydrogen 50%

THE RATIO OF GASES in the atmosphere is shown in the chart at left. Clouds form in the troposphere; the overlying stratosphere, extending 50 miles above the earth, is clear. The ionosphere (50-200 miles) contains layers of charged particles (ions) that reflect radio waves, permitting messages to be transmitted over long distances. Faint traces of atmosphere exist in the exosphere to about 500 miles from the earth's surface.

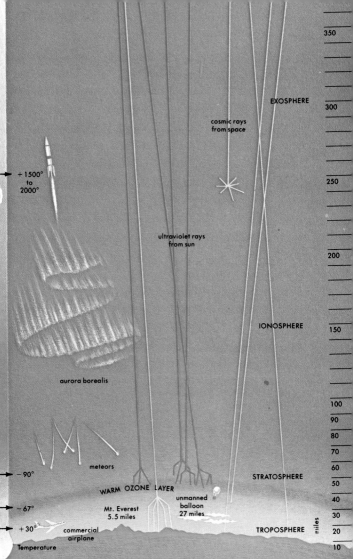

350

EXOSPHERE

300

cosmic rays
from space

+1500°
to
2000°

250

ultraviolet rays
from sun

200

IONOSPHERE

150

aurora borealis

100
90
80
70
60

meteors

STRATOSPHERE

50

−90°

WARM OZONE LAYER

40

−67°

Mt. Everest
5.5 miles

unmanned
balloon
27 miles

30

+30°

commercial
airplane

TROPOSPHERE

20

miles

10

Temperature

0

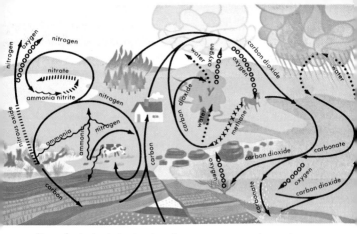

Atmospheric circulation involves the continuous recirculation of various substances.

OUR PRESENT ATMOSPHERE and oceans were probably derived by degassing of the semi-molten earth and continuing later additions from volcanoes and hot springs. These gases—such as hydrogen, nitrogen, hydrogen chlorides, carbon monoxide, carbon dioxide, and water vapor—probably formed the atmosphere of earlier geologic times. The lighter gases, such as hydrogen, probably escaped. The later development of living organisms capable of photosynthesis slowly added oxygen to the atmosphere, ultimately allowing the colonization of the land by providing free oxygen for respiration and also forming the ozone layer, which shields the earth from ultraviolet radiation of the sun.

Some evidence for this sequence in the development of the atmosphere is contained in the sequence of Precambrian rocks and fossils, which suggests a transition from a non-oxygen to free-oxygen environment.

THE EARTH'S CRUST: COMPOSITION

We have so far been able to penetrate to only very shallow depths beneath the surface of the earth. The deepest mine is only about 2 miles deep, and the deepest well about 5 miles deep.

But by using geophysical methods we can "x-ray" the earth. Careful tracing of earthquake waves shows that the earth has a distinctly layered structure. Studies of rock density and composition, heat flow, and magnetic and gravitational fields also aid in constructing an earth model of three layers: crust, mantle, and core. Estimates of the thickness of these layers, and suggested physical and chemical characteristics form an important part of modern theories of the earth (p. 126).

The crust of the earth is formed of many different kinds of rocks (p. 92), each of which is an aggregate of minerals, described on pp. 22-31.

Grand Canyon of Colorado River, Arizona, is 1 mile deep, but exposes only a small part of upper portion of earth's crust.

MINERALS are naturally occurring substances with a characteristic atomic structure and characteristic chemical and physical properties. Some minerals have a fixed chemical composition; others vary within certain limits. It is their atomic structure that distinguishes minerals from one another.

Some minerals consist of a single element, but most minerals are composed of two or more elements. A diamond, for instance, consists only of carbon atoms, but quartz is a compound of silica and oxygen. Of the 105 elements presently known, nine make up more than 99 percent of the minerals and rocks.

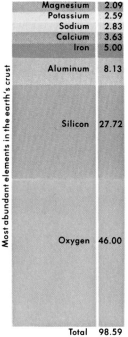

Most abundant elements in the earth's crust	
Magnesium	2.09
Potassium	2.59
Sodium	2.83
Calcium	3.63
Iron	5.00
Aluminum	8.13
Silicon	27.72
Oxygen	46.00
Total	98.59

OXYGEN AND SILICON are the two most abundant elements in the earth's crust. Their presence, in such enormous quantities, indicates that most of the minerals are silicates (compounds of metals with silicon and oxygen) or aluminosilicates. Their presence in rocks is also an indication of the abundance of quartz (SiO_2, silicon dioxide) in sandstones and granites, as well as in quartz veins and geodes.

Smoky Quartz

The most striking feature of minerals is their crystal form, and this is a reflection of their atomic structure. The simplest example of this is rock salt, or halite (NaCl, sodium chloride), in which the positive ions (charged atoms) of sodium are linked with negatively charged chlorine ions by their unlike electrical charges. We can imagine these ions as spheres, with the spheres of sodium having about half the radius of the chlorine ions (.98 Å as against 1.8 Å; Å is an Ångstrom Unit, which is equivalent to one hundred millionth of a centimeter, written numerically as 0.00000001 cm or 10^{-8} cm). The unit is named for Anders Ångstrom, a Swedish physicist.

SODIUM ATOM joins **CHLORINE ATOM**

X-RAY STUDIES show that the internal arrangement of halite is a definite cubic pattern, in which ions of sodium alternate with those of chlorine. Each sodium ion is thus held in the center of and at equal distance from six symmetrically arranged chlorine ions, and vice versa. It is this basic atomic arrangement or crystalline structure that gives halite its distinctive cubic crystal form and its characteristic physical properties.

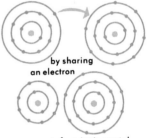

by sharing
an electron

to form ionic crystal
sodium chloride

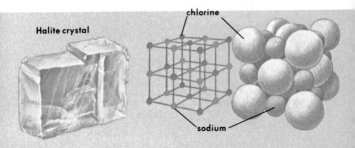

Halite crystal

chlorine

sodium

THE ATOMIC STRUCTURE of each mineral is distinctive but most minerals are more complicated than halite, some because they comprise more elements, others because the ions are linked together in more complex ways. A good example of this is the difference between diamond and graphite. Both have an identical chemical composition (they are both pure carbon) but they have very different physical properties. Diamond is the hardest mineral known, and graphite is one of the softest. Their different atomic structures reflect their different geologic modes of origin.

DIAMOND, the hardest natural substance known, consists of pure carbon atoms. Each carbon atom is linked with four others by electron-sharing. The four electrons in the outer shell are shared with four neighboring atoms. Each atom of carbon then has eight electrons in its outer shell. This provides a very strong bond. Its crystal form is a reflection of its structure and of the conditions under which it was formed. Diamond is usually pale yellow or colorless, but is found also in shades of red, orange, green, blue, brown, or black. Pure white or blue-white are best for gems.

GRAPHITE, quite different from diamond, is soft and greasy, and widely used as an industrial lubricant. In graphite, carbon atoms are arranged in layers, giving the mineral its flaky form. The atoms within each layer have very strong bonds, but those that hold successive layers together are very weak. Some atoms between layers are held together so poorly that they move freely, giving the graphite its soft, slippery, lubricating properties. Because of its poor bonding, graphite is a good conductor of electricity. Its best-known use is in "lead" pencils.

CRYSTAL FORM of minerals is an important factor in their identification. Grown without obstruction, minerals develop a characteristic crystal form. The outer arrangement of plane surfaces reflects their internal structure. Perfect crystals are rare. Most minerals occur in irregular masses of small crystals because of restricted growth. Since all crystals are three-dimensional, they may be classified on the basis of the intersection of their axes. Axes are imaginary lines passing through the geometric center of a crystal from the middle of its faces and intercepting in a single point.

CUBIC CRYSTALS have three axes of equal length meeting at right angles, as in galena, garnet, pyrite, and halite.

TETRAGONAL CRYSTALS have three axes at right angles, two of equal length, as in zircon, rutile, and scapolite.

Galena   Zircon

HEXAGONAL CRYSTALS have three equal horizontal axes with 60° angles and one shorter or longer at right angles, as in quartz and tourmaline.

ORTHORHOMBIC CRYSTALS have three axes at right angles, but each is of different length, as in barite and staurolite.

Quartz   Staurolite

MONOCLINIC CRYSTALS have three unequal axes, two forming an oblique angle and one perpendicular, as in augite, orthoclase, and epidote.

TRICLINIC CRYSTALS have three axes of unequal lengths, none forming a right angle with others, as in plagioclase feldspars.

Epidote   Amazonite feldspar

MINERAL IDENTIFICATION involves the use of various chemical and physical tests to determine what minerals are present in rock. There are over 2,000 minerals known, and elaborate laboratory tests (such as X-ray diffraction) are required to identify some of them. But many of the common minerals can be recognized after a few simple tests. Six important physical properties of minerals (hardness, luster, color, specific gravity, cleavage, and fracture) are easily determined. A balance is needed to find specific gravity of crystals or mineral fragments. For the other tests, a hand lens, steel file, knife, and a few other common items are helpful. Specimens can sometimes be recognized by taste, tenacity, tarnish, transparency, iridescence, odor, or the color of their powder streak, especially when these observations are combined with tests for the other physical properties.

HARDNESS is the resistance of a mineral surface to scratching. Ten well-known minerals have been arranged in a scale of increasing hardness (Mohs' scale). Other minerals are assigned comparable numbers from 1 to 10 to represent relative hardness. A mineral that scratches orthoclase (6) but is scratched by quartz (7) would be assigned a hardness value of 6.5.

LUSTER is the appearance of a mineral when light is reflected from its surface. Quartz is usually glassy; galena, metallic.

MOHS' SCALE OF HARDNESS

	10	diamond
	9	corundum
	8	topaz
	7	quartz
steel file	6	orthoclase
glass	5	apatite
knife blade	4	fluorite
penny	3	calcite
fingernail	2	gypsum
	1	talc

Galena crystals

SPECIFIC GRAVITY is the relative weight of a mineral compared with the weight of an equal volume of water. A balance is normally used to determine the two weights. Some minerals are similar superficially but differ in density. Barite may resemble quartz, but quartz has a specific gravity of 2.7; barite, 4.5.

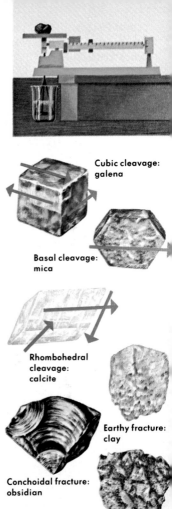

Cubic cleavage: galena

Basal cleavage: mica

Rhombohedral cleavage: calcite

Conchoidal fracture: obsidian

Earthy fracture: clay

Uneven fracture: arsenopyrite

COLOR varies in some minerals. Pigments or impurities may be the cause. Quartz occurs in many hues but is sometimes colorless. Among minerals with a constant color are galena (lead gray), sulfur (yellow), azurite (blue), and malachite (green). A fresh surface is used for identification, as weathering changes the true color.

CLEAVAGE is the tendency of some minerals to split along certain planes that are parallel to their crystal faces. A hammer blow or pressure with a knife blade can cleave a mineral. Galena and halite have cubic cleavage. Mica can be separated so easily that it is said to have perfect basal cleavage. Minerals without an orderly internal arrangement of atoms have no cleavage.

FRACTURE is the way a mineral breaks other than by cleavage. Minerals with little or no cleavage are apt to show good fracture surfaces when shattered by a hammer blow. Quartz has a shell-like fracture surface. Copper has a rough, hackly surface; clay, an earthy fracture.

27

COMMON ROCK-FORMING MINERALS include carbonates, sulfates, and other compounds. Many minerals crystallize from molten rock material. A few form in hot springs and geysers, and some during metamorphism. Others are formed by precipitation, by the secretions of organisms, by evaporation of saline waters, and by the action of ground water.

MINERAL CARBONATES, SULFATES, AND OXIDES

LIMONITE is a group name for hydrated ferric oxide minerals, $Fe_2O_3.H_2O$. It is an amorphous mineral that occurs in compact, smooth, rounded masses or in soft, earthy masses. No cleavage. Earthy fracture. Hardness (H) 5 to 5.5; Sp. Gr. 3.5 to 4.0. Rusty or blackish color. Dull, earthy luster gives a yellow-brown streak. Common weathering product of iron minerals.

CALCITE is a calcium carbonate, $CaCO_3$. It has dogtooth or flat hexagonal crystals with excellent cleavage. H. 3; Sp. Gr. 2.72. Colorless or white. Impurities show colors of yellow, orange,

GYPSUM is a hydrated calcium sulphate, $CaSO_4.2H_2O$. Tabular or fibrous monoclinic crystals, or massive. Good cleavage. H. 2. Sp. Gr. 2.3. Colorless or white. Vitreous to pearly luster. Streaks are white. Flexible but no elastic flakes. Sometimes fibrous. Found in sedimentary evaporites and as single crystals in black shales. The compact, massive form is known as alabaster.

brown, and green. Transparent to opaque. Vitreous or dull luster. Major constituent of limestone. Common cave and vein deposit. Reacts strongly in dilute hydrochloric acid.

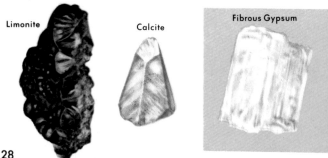

Limonite

Calcite

Fibrous Gypsum

QUARTZ is silicon dioxide, SiO_2. Massive or prismatic. No cleavage. Conchoidal fracture. H. 7; Sp. Gr. 2.65. Commonly colorless or white. Vitreous to greasy luster. Transparent to opaque. Common in acid igneous, metamorphic, and clastic rocks, veins, and geodes. The most common of all minerals.

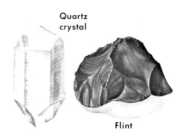

Quartz crystal

Flint

ROCK-FORMING SILICATE MINERALS

FELDSPARS are alumino-silicates of either potassium ($KAlSi_3O_8$ orthoclase, microcline, etc.) or sodium and calcium (plagioclase feldspars $NaAlSi_3O_8$, $CaAl_2Si_2O_8$). Well-formed monoclinic or triclinic crystals, with good cleavage. H. 6 to 6.5; Sp. Gr. 2.5 to 2.7 Orthoclase feldspars are white, gray, or pink, vitreous to pearly luster, and lack surface striations. Plagioclase feldspars are white or gray, have two good cleavages, which produce fine parallel striations on cleavage surfaces. Common in igneous and metamorphic rocks, and arkosic sandstones.

MICAS are silicate minerals. White mica (muscovite) is a potassium alumino-silicate. Black mica (biotite) is a potassium, iron, magnesium alumino-silicate. Both occur in thin, monoclinic, pseudo-hexagonal, scalelike crystals. Superb cleavage gives thin, flexible flakes. Pearly to vitreous luster. Micas are common in igneous, metamorphic, and sedimentary rocks.

Microline

Albite (plagioclase feldspar)

Biotite crystal

Biotite (black mica)

Augite

PYROXENES include a large group of silicates of calcium, magnesium, and iron. Augite, $(CaMgFeAl)_2 \cdot (AlSi)_2O_6$, and hypersthene, $(FeMg)SiO_3$, are the most common. Stubby, eight-sided prismatic, orthorhombic or monoclinic crystals, or massive. Two cleavages meet at 90° (compare amphiboles), but these are not always developed. Gray or green, grading into black. Vitreous to dull luster. H. 5 to 6. Sp. Gr. 3.2 to 3.6. Common in nearly all basic igneous and metamorphic rocks. Sometimes found in meteorites.

Hornblende

AMPHIBOLES are complex hydrated silicates of calcium, magnesium, iron, and aluminum. Hornblende, a common amphibole, has long, slender, prismatic, six-sided orthorhombic or monoclinic crystals; sometimes fibrous. Two good cleavages meeting at 56°. H. 5 to 6; Sp. Gr. 2.9 to 3.2. Black or dark green. Opaque with a vitreous luster. Common in basic igneous and metamorphic rocks. Asbestos is an amphibole.

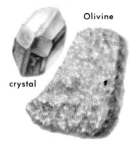

Olivine

crystal

OLIVINE is a magnesium–iron silicate, $(FeMg)_2SiO_4$. Small, glassy grains. Often found in large, granular masses. Crystals are relatively rare. Poor cleavage. Conchoidal fracture. H. 6.5 to 7; Sp. Gr. 3.2 to 3.6. Various shades of green; sometimes yellowish. Transparent or translucent. Vitreous luster. Common in basic igneous and metamorphic rocks. Olivine alters to a brown color.

COMMON ORE MINERALS

GALENA is a lead sulphide, PbS. Heavy, brittle, granular masses of cubic crystals. Perfect cubic cleavage, H. 2.5; Sp. Gr. 7.3 to 7.6 Silver-gray. Metallic luster. Streaks are lead-gray. Important lead ore. Common vein mineral. Occurs with zinc, copper, and silver.

SPHALERITE is a zinc sulphide, ZnS. Cubic crystals or granular, compact. Six perfect cleavages at 60°. H. 3.5 to 4; Sp. Gr. 3.9 to 4.2. Usually brownish; sometimes yellow or black. Translucent to opaque. Resinous luster. Some specimens are fluorescent. Important zinc ore. Common vein mineral with galena.

PYRITE is an iron sulphide, FeS_2. Cubic, brassy crystals with striated faces. May be granular. No cleavage. Uneven fracture. H. 6 to 6.5; Sp. Gr. 4.9 to 5.2. Brassy yellow color. Metallic luster. Opaque and brittle. Also called fool's gold. Common source of sulfur.

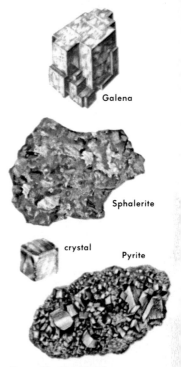

Galena

Sphalerite

crystal

Pyrite

Miner in the lead mine Mesters Vig, Greenland

THE CRUST: EROSION AND DEPOSITION

The earth's crust is influenced by three great processes which act together:

Gradation includes the various surface agencies (in contrast to the two internal processes below), which break down the crust (degradation) or build it up (aggradation). Gradation is brought about by running water, winds, ice and the oceans. Most sediments are finally deposited in the seas.

Diastrophism is the name given to all movements of the solid crust with respect to other parts (p. 106). Sometimes this involves the gentle uplift of the crust. Many rocks that were formed as marine sediments gradually rose until they now stand thousands of feet above sea level. Other diastrophic movements may involve intensive folding and fracture of rocks.

COASTLINES the world over provide evidence of changes in the earth's unstable crust. The photograph shows cliffs made of rocks that were deposited under the seas that covered the area about 130 million years ago. These were uplifted and folded, so their original layers now stand almost vertical. At present they are undergoing erosion by the sea, typified by the form of the arch. Eroded material is being redeposited as a beach. The rocks of which coastlines are formed are themselves the result of earlier gradational events.

NATURAL ARCH reflects process of erosion, while deposition has produced the beach. Dorset, England.

Eruption of Kapoho, Hawaii, showing paths of molten lava

VULCANISM includes all the processes associated with the movement of molten rock material. This includes not only volcanic eruptions but also the deep-seated intrusion of granites and other rocks (p. 83).

These three processes act so that at any time the form and position of the crust is the result of a dynamic equilibrium between them, always reflecting the climate, season, altitude, and geologic environment of particular areas. As an end product of degradation, the continents would be reduced to flat plains, but the balance is restored and the processes of erosion counteracted by other forces that tend to elevate parts of the earth's crust. These changes reflect changes in the earth's interior (p. 146).

Yosemite valley is the result of interaction of various types of erosional processes.

EROSION involves the breaking down and removal of material by various processes or degradation.

YOSEMITE VALLEY in California is a good example of the complex interplay of gradational processes. A narrow canyon was first carved by the River Merced. This was later deepened and widened by glaciation. Running water is now modifying the resultant hanging and U-shaped valleys (p.

58), so characteristic of glacial topography. The level of the main valley floor lies 3,000 feet below the upland surface of the Sierras. Differences in topography are partly the result of differences in jointing and resistance of underlying granites. Half Dome and El Capitan are resistant granitic monoliths laid bare by this differential weathering. The region thus shows the effect of many degradational processes. But the 300-ft.-thick sediment on the valley floor reveals the continuing aggradational effects that are also at work.

WEATHERING

Weathering is the general name for all the ways in which a rock may be broken down. It takes place because minerals formed in a particular way (say at a high temperature in the case of an igneous rock) are often unstable when exposed to the various conditions affecting the crust of the earth. Because weathering involves interaction of the lithosphere with the atmosphere and hydrosphere, it varies with the climate. But all kinds of weathering ultimately produce broken mineral and rock fragments and other products of decomposition. Some of these remain in one place (clay or laterite, for example) while others are dissolved and removed by running water.

The earth's surface, above the level of the water table (p. 50), is everywhere subject to weathering. The weathered cover of loose rock debris (as opposed to solid bedrock) is known as the *regolith*. The thickness and distribution of the regolith depend upon both the rate of weathering and the rate of removal and transportation of weathered material.

THE EFFECTS of weathering are most strikingly seen in arid and semiarid environments, where bare rocks are exposed without a cover of vegetation. Bryce Canyon, Utah, shows the effects of the bedding and differing resistance of rocks in producing distinctive erosional landforms. Weathering is of great importance to humankind. Soils are the result of weathering processes, and are enriched by the activities of animals and plants. Some important economic resources, such as our ores of iron and aluminum, are the result of residual weathering processes.

FROST SHATTERING is often produced by alternate freezing and thawing of water in rock pores and fissures. Expansion of water during freezing causes the rock to fracture.

SPHEROIDAL WEATHERING occurs in well-jointed rocks, because weathering takes place more rapidly at corners and edges (3 and 2 sides) than on single faces.

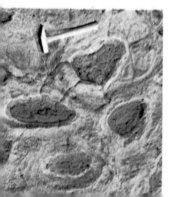

MECHANICAL WEATHERING involves the disintegration of a rock by mechanical processes. These include freezing and thawing of water in rock crevices, disruption by plant roots or burrowing animals, and the changes in volume that result from chemical weathering within the rock. This weathering is especially common in high latitudes and altitudes, which have daily freezing and thawing, and in deserts, where there is little water or vegetation. Rather angular rock forms are produced, and little chemical change in the rock is involved. It was once thought that extreme daily temperature changes caused mechanical weathering, but this now seems uncertain.

CHEMICAL WEATHERING involves the decomposition of rock by chemical changes or solution. The chief processes are oxidation, carbonation and hydration, and solution in water above and below the surface. Many iron minerals, for example, are rapidly oxidized ("rusted") and limestone is dissolved by water containing carbon dioxide. Such decomposition is encouraged by warm, wet climatic conditions and is most active in tropical and temperate climates. Blankets of soil or other material are produced which are so thick and extensive that solid rock is rarely seen in the tropics. Chemical weathering is more widespread and common than mechanical weathering, although usually both act together.

SOIL is the most obvious result of weathering. It is the weathered part of the crust capable of supporting plant life. The thickness and character of soil depend upon rock type, relief, climate, and the "age" of a soil, as well as the effect of living organisms.

Immature soils are little more than broken rock fragments, grading down into solid rock. Mature soils include quantities of humus, formed from decayed plants, so that the upper surface (topsoil) becomes dark. Organic acids and carbon dioxide released during vegetative decay dissolve lime, iron, and other compounds and carry them down into the lighter subsoil.

Residual soils, formed in place from the weathering of underlying rock, include laterites, produced by tropical leaching and oxidizing conditions which consist of iron and aluminum oxides with almost no humus. Transported soils have been carried from the parent rocks from which they formed and deposited elsewhere. Wind-blown loess (p. 77), alluvial deposits (p. 44), and glacial till (p. 59) are common examples of transported soils.

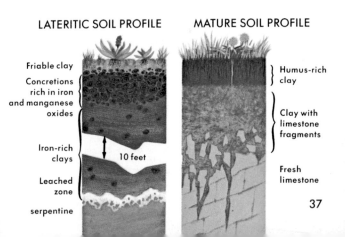

LATERITIC SOIL PROFILE

Friable clay
Concretions rich in iron and manganese oxides
Iron-rich clays — 10 feet
Leached zone
serpentine

MATURE SOIL PROFILE

Humus-rich clay
Clay with limestone fragments
Fresh limestone

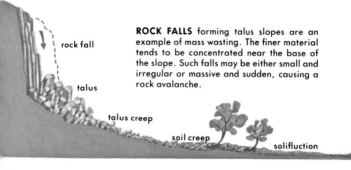

ROCK FALLS forming talus slopes are an example of mass wasting. The finer material tends to be concentrated near the base of the slope. Such falls may be either small and irregular or massive and sudden, causing a rock avalanche.

rock fall

talus

talus creep

soil creep

solifluction

MASS WASTING is the name given to all downslope movements of regolith under the predominant influence of gravity. Weathered material is transported from its place of origin by gravity, streams, winds, glaciers, and ocean currents. Each of these agencies is a depositing, as well as a transporting agent and, though they rarely act independently, each produces rather different results.

The prevention of mass wasting of soil is of great importance in all parts of the world. Engineering and mining activities usually require geological advice on slide and subsidence dangers. Several tragic dam failures have resulted from slides. The Vaiont reservoir slide in Italy in 1963 claimed 2,600 lives. Careful geological site surveys can prevent such disasters.

EARTH FLOWS may be slow or very rapid. Slow movement (solifluction) takes place in areas where part of the ground is permanently frozen. These areas cover about 20 percent of the earth. On slopes, the thawed upper layer slides on the frozen ground below it. On flat ground, lateral movement gives stone polygons. Sudden flows may follow heavy rains.

SOIL CREEP is the gentle, downhill movement of the regolith that occurs even on rather moderate, grass-covered slopes. It can often be seen in road cuts.

SLUMPS are slides in soft, unconsolidated sediment. Some are submarine in origin. Fossil slumps are recognizable in some strata. Usually, slumps are on a rather small scale, as when sod breaks off a stream bank.

Hill creep affects fences and trees, as well as bending of vertical strata.

SUBSIDENCE is downward movement of the earth's surface caused by natural means (chemical weathering in limestone areas) or artificial methods (excessive mining or brine pumping). Heavy loading by buildings or engineering structures may also cause subsidence.

LANDSLIDES are mass movements of earth or rock along a definite plane. They occur in areas of high relief where weak planes—bedding, joints, or faults (p. 110-111)—are steeply inclined; where weak rocks underlie massive ones; where large rocks are undercut; and where water has lubricated slide planes.

Typical landslide topography in Wyoming shows debris of huge boulders.

Rock glaciers involve slow downslope movement of "river" of rock.

RUNNING WATER

Running water is the most powerful agent of erosion. Wind, glaciers, and ocean waves are all confined to relatively limited land areas, but running water acts almost everywhere, even in deserts. One fourth of the 35,000 cubic miles of water falling on the continents each year runs off into rivers, carrying away rock fragments with it. In the United States, this erosion of the land surface takes place at an average rate of about one inch in 750 years. Running water breaks down the crust by the impact of rock debris it carries.

THE HYDROLOGIC CYCLE is a continuous change in the state of water as it passes through a cycle of evaporation, condensation, and precipitation. Of the water that falls on the land, up to 90 percent is evaporated. Some is absorbed by plants and subsequently transpired to the atmosphere, some runs off in streams and rivers, and some soaks into the ground. The relative amounts of water following these paths vary considerably and depend upon the slope of the ground; the character of the soil and rocks; the amount, rate, and distribution of rainfall; the amount and type of plant cover; and the temperature. The hydrosphere is estimated to include over 300 million cubic miles of water, about 97 percent of which is in the oceans.

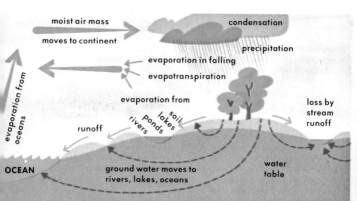

moist air mass moves to continent

condensation

precipitation

evaporation in falling

evapotranspiration

evaporation from soil, lakes, ponds, rivers

evaporation from oceans

loss by stream runoff

runoff

OCEAN

ground water moves to rivers, lakes, oceans

water table

A single shower may involve the downpour of more than a billion tons of water. Each raindrop in the shower becomes important in erosion, especially in areas with sparse vegetation and unconsolidated sediments. Runoff water rarely travels far as a continuous sheet, for it is broken up into rivulets and streams by surface irregularities in rock type and relief.

Running water carries its load of rock debris partly in suspension, partly by rolling and bouncing it along the bottom, and partly in solution. The carrying power of a stream is proportional to the square of its velocity, and so is enormously increased in time of flood.

THE FORM OF RIVERS depends upon many factors. Water runs downhill under the influence of gravity, the flow of the water being characteristically turbulent, with swirls and eddies. The overall long profile of a river valley is concave upwards, however much its gradient (its slope or fall in a given distance) may vary. The velocity of the stream increases with the gradient, but also depends on other factors, including the position within the river channel, the degree of turbulence, the shape and course of the channel, and the stream load (transported materials).

Diagrammatic model of a river system, showing cross sections of channel at three points

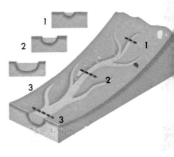

DENDRITIC DRAINAGE PATTERN of Diamantina River, Queensland, Australia, is typical of river development in areas where the underlying rocks are relatively uniform in their resistance to erosion. This pattern may be modified by continued downcutting of the river.

Niagara Falls are formed by resistant bed of dolomite.

RIVER PROFILES reflect the varying development of drainage systems. During early development, changes are reflected by direction and shape of stream courses. Ultimately, erosion produces an equilibrium between the slope and volume of the river and its erosional and depositional power. This results in a graded profile, or profile of equilibrium. Final equilibrium is never reached because of seasonal and geologic changes. The downward limit of erosion by a river is the base level, below which the river cannot downcut appreciably because it has reached the level of the body of water into which it flows. Sea level is the ultimate base level for all rivers. Larger rivers and lakes into which rivers flow constitute local base levels.

Stages in the cycle of river erosion were once labeled as "youth," "maturity," and "old age." Although these stages describe certain characteristics, they imply no particular age in years, only changing phases in development. Rivers often show a condition of old age near their mouths, but are mature or youthful in their higher reaches. For that reason, the terms are rather misleading, and are now rarely used.

RIVER PROFILES are influenced by geology and structure of their drainage area and tend to assume a graded, concave profile as they approach equilibrium.

IRREGULAR PROFILE—common in upstream portions of rivers—shows high gradients, waterfalls, rapids, steep-sided valleys, irregular courses, few tributaries, and erosion. The Gunnison River in Colorado is an example.

Longitudinal profile ▶

SMOOTHER STEEP PROFILE is steep, with high relief, but fewer irregularities. Erosion and transportation of underlying rocks gives wider valleys and smoother topography. Tributaries are well-established. The White River of southwestern Missouri shows these features.

Longitudinal profile ▶

GRADED PROFILE reflects deposition of transported load. The river flows sluggishly over a wide, flat flood plain, with meandering pattern, ox-bow lakes, and sand bars. This represents a baseline equilibrium between erosion by the river and deposition in the sea or lake into which it flows. The lower Mississippi and Amazon are examples.

Longitudinal profile ▶

REJUVENATION or uplift may interrupt the cycle of erosion at any stage to provide new energy for downcutting. The character of the stream is then often a combination of recently cut, steep gorges in an older meander course. Ancient flood-plains are often left "stranded" as terraces. Uplift and warping may be relatively sudden or slow, frequent or rare, local or regional in extent.

Terraces are cut as river swings from one side of valley to other.

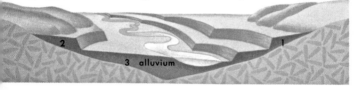

2 1 3 alluvium

WATERFALLS AND RAPIDS are local increases in gradient in the long profile of a river. Most are due to unequal erosion of the stream bed.

THE FALLS of the Yellowstone River (the lower of which is twice as high as Niagara Falls) result from resistant lava flows. Niagara Falls is held up by an 80-foot-thick bed of dolomite, which is more resistant to erosion than the underlying shales and thin limestones.

Other falls result from over-deepening of a main valley, often by ice, as in the Bridal Veil Falls in Yosemite, so that the tributaries are left "hanging" (p. 34). Continued erosion by a stream leads to upstream migration and ultimate smoothing out of waterfalls.

UNEQUAL EROSION has already cut Niagara Falls back about 7 miles from the Niagara escarpment, since it began cutting about 9,000 years ago. The weaker shales below the Lockport Dolomite are undercut by the turbulent water in the plunge pool, so that the overlapping dolomite is undermined, and eventually collapses. The Niagara River flows into Lake Ontario from Lake Erie, which will ultimately be drained by upstream migration of the falls.

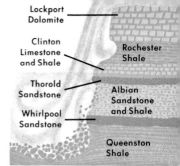

Lockport Dolomite

Clinton Limestone and Shale

Rochester Shale

Thorold Sandstone

Albian Sandstone and Shale

Whirlpool Sandstone

Queenston Shale

STREAM PIRACY takes place when the tributaries of one stream (A) erode faster than those of another (B) in the same area. The headwaters of the less active stream are ultimately diverted or cut off. When such "beheaded" streams abandon their path through a ridge, a wind gap results.

headward erosion

B

A

DELTAS are masses of sediment deposited where rivers lose their velocity as they enter lakes or seas. Thick, relatively uniform layers of sediment accumulate on the steep outward slope. Deltas of the Mississippi, Ganges, and Po are thousands of square miles in area. Deltas have a characteristic stratification in their deposits.

▼

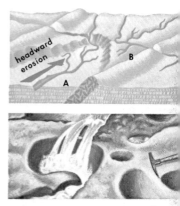

POTHOLES are circular hollows ▲ in a stream bed, drilled out by swirling currents of water carrying gravel and pebbles. This "hydraulic drilling" is an important method of down-cutting, even in hard rock.

Coarse-grained sediments

Fore-set beds

Fine-grained sediments

45

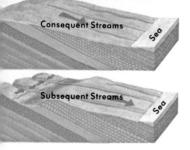

DRAINAGE

CONSEQUENT STREAMS flow in directions determined largely by the original slope and shape of the ground. When their courses are modified by features of the geology, such as valley cutting in soft strata, the adjusted tributaries are known as subsequent streams.

DENDRITIC drainage patterns are those that show treelike branching because the bedrock has a uniform resistance to erosion and does not influence the direction of stream flow.

TRELLIS patterns are characteristic of uniformly dipping or strongly folded rocks. In this rectangular pattern, the tributaries are nearly perpendicular to the main stream.

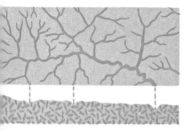

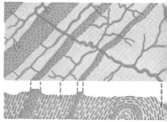

SUPERIMPOSED drainage generally has stream courses that are independent of rock structure. A stream's erosional power may have been strong enough to maintain its antecedent course during

uplift and development of a new geological structure. Or it may keep its same course after it cuts through younger, overlying, flat, sedimentary rock to an older, irregular rock mass.

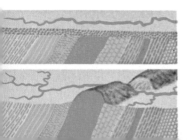

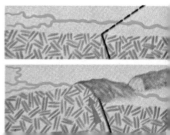

PATTERNS

RADIAL patterns develop on young mountains, such as volcanoes where streams radiate from the high central area.

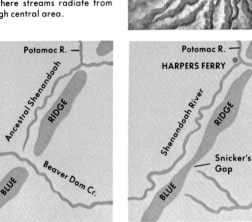

Piracy may cause river diversion. Ancestral Shenandoah River captured headwaters of Beaver Dam, leaving an abandoned water gap.

MODIFIED DRAINAGE may be caused by piracy (p. 45) as well as by glacial or volcanic blocking of stream courses. Glacial diversion results from overdeepening of basins and river valleys by ice, blocking of drainage by ice, morainic deposits, and meltwater, which may produce glacial lakes and new outlets. Changes in sea level may also modify drainage.

Great Lakes basins were carved by ice from soft strata. Original drainage was blocked by ice, producing local crustal depressions.

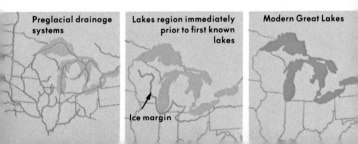

Preglacial drainage systems

Lakes region immediately prior to first known lakes

Ice margin

Modern Great Lakes

EROSIONAL LANDFORMS are produced by running water and other erosional agents. *Mesas* are flat-topped rock mountains, which stand as remnants of a once continuous plateau. *Buttes* are smaller examples of the same thing. *Monuments* describe any isolated rock pinnacle.

hogback escarpment

cuesta   mesa   butte

HOGBACKS are long ridges formed by steeply dipping resistant strata; cuestas are gently sloping ridges formed in gently dipping strata.

NATURAL BRIDGES are formed of resistant strata, usually sandstone or limestone. Underground erosion has taken place below the original stream bed.

◄ **PINNACLES** at Bryce Canyon, Utah, show differential weathering. Erosion has removed the softer, more soluble rocks. Rocks here are of Tertiary (Eocene) age.

GROUNDWATER is found almost everywhere below the earth's surface. Most originates from rain and snow, but small quantities come from water trapped in sediments during their deposition (connate water) or from igneous magmas (juvenile water).

SEPTARIAN NODULE is formed from minerals deposited by groundwater in claylike rocks. Groundwater is an important agent in both deposition and erosion of surface rocks. It can dissolve original cementing materials and deposit new ones. For example, it produces caves and caverns by solution in carbonate rocks.

POROSITY, the percentage of pore space to total volume of a rock, depends upon the grain size, shape, packing, and cement of rock particles. The permeability of a rock, or its capacity to transmit or yield water, depends upon the size of pores, rather than their total volume. Pores smaller than 1/20 of a millimeter will not allow water to flow through them.

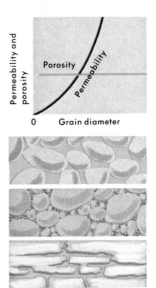

WELL-SORTED SAND (enlarged view) with high porosity due to sorting, which has removed fine-grained particles

POORLY SORTED SAND has lower porosity than well-sorted sand, because the pore spaces are filled by fine particles.

LIMESTONES generally hold water in enlarged joints formed by solution; they lack the "pore space" of sandstones.

49

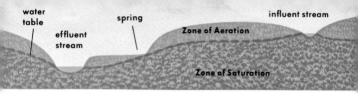

THE WATER TABLE depends on the distribution of groundwater. The open spaces in the rocks of the upper part of the crust are filled mainly with air. This is the zone of aeration. Water moves downward through this zone into the zone of saturation, where openings are filled with water. The upper surface of this saturated zone is the water table. In most areas, the water table is only a few tens of feet below the surface, but in arid regions, it is much deeper. Water-bearing rocks are rarely found below 2,000 feet. Rock pores are closed by pressure at depth, and this determines the lower limit for groundwater. Most rocks will give off water whenever they intersect the water table. But the level of the water table falls after a dry season, so rock formations must be deep enough to penetrate the water table all year round. Sometimes a perched water table results when a pocket of water is held above the normal water table by a saucer of impervious rock. Any water-producing rock formation is called an aquifer.

ARTESIAN WELLS are those where water is confined to a permeable aquifer by impervious beds, and where the catchment or intake area (and thus the water level in the aquifer) is higher than the well head. This allows the water to flow toward the surface under its own internal pressure.

water table

shale

porous sandstone

impermeable shale

50

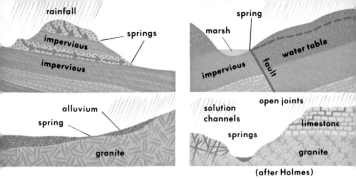

(after Holmes)

SPRINGS are sources of running water produced by the water table intersecting the ground surface. A few of the many ways they can be formed are shown in the diagrams above. Some springs are dry at seasons when the water table is depressed; others flow without interruption.

HOT SPRINGS are generally confined to areas of recent vulcanism where groundwater is heated at depth by contact with igneous magmas. Such springs are well developed in Yellowstone National Park and North Island, New Zealand. Terrace deposits may be produced when hot spring water deposits dissolve mineral matter. Mammoth Springs of Yellowstone National Park are formed of calcium carbonate (travertine). Geysers, intermittent fountainlike hot springs, often build cones of siliceous geyserite.

Hot Springs, Thermopolis, Wyoming

Norris Geyser Basin, Wyoming

FUMAROLES are gentle geysers located in volcanic regions that emit fumes, usually in the form of steam.

GEYSERS are thermal springs that periodically discharge their water with explosive violence. All geysers have a long narrow pipe extending down from their vents into their reservoirs. A build-up and sudden release of steam bubbles probably relieves the pressure on the heated water below ground so that it boils and surges upward. The period between eruptions varies from minutes to months in different geysers, depending upon the structure of the geyser, its water supply, and its heat source.

OLD FAITHFUL in Yellowstone National Park discharges a column of water and steam up to 170 feet high approximately every 65 minutes.

Typical geyser structure shows complex system of fissures extending down to regions where ground water becomes superheated and finally erupts.

Pudding Basin Geyser, New Zealand shows typical eruption.

▼

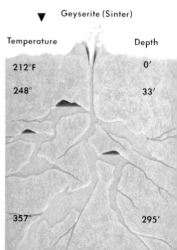

Geyserite (Sinter)

▼

Temperature	Depth
212°F	0'
248°	33'
357°	295'

CAVES reflect the work of groundwater. Limestone is dissolved by circulating water in subsurface joints and fissures. The enlargement gradually produces a cave. Inside a cave, dripping water, rich in calcium bicarbonate and CO_2, often produces precipitates that form stalagmites and stalactites.

GEOLOGICAL WORK OF GROUNDWATER is important in solution and deposition in the rocks through which it passes. It dissolves limestone and other carbonate rocks to form caves and sink holes. Limestone areas are often marked by a karst topography of sinks, few surface streams (they flow underground), and large springs. In caves, the deposition of calcite dissolved in dripping groundwater produces stalactites, icicle-shaped formations hanging from the cave ceilings, and stalagmites, formations that build up from the cave floor. The carrying away of minerals in solution usually occurs above the water table. Below that level, deposition, replacement, and cementation are important. Ball-like masses (concretions), hollow, globular bodies (geodes), and the cement in many sedimentary rocks are the result of the action of groundwater at depth.

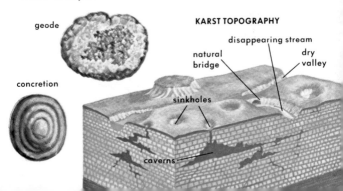

geode

concretion

KARST TOPOGRAPHY

disappearing stream

natural bridge

dry valley

sinkholes

caverns

Cotter Dam supplies water for Australian capital city of Canberra.

THE EARTH'S WATER SUPPLY is one of its most precious natural resources. Although nearly three quarters of the globe is covered by water, over 97 percent of the 326 million cubic miles of earth's water is locked up in the oceans, too salty for drinking water or for agriculture. Another 2 percent is frozen in glaciers and ice sheets. The tiny fraction that is available for water supply is very unevenly distributed. One third of the earth's land surface is desert or semiarid. Even in humid areas, water supply and conservation present major problems. The location and development of new industries depend upon adequate water supplies. World demand for water is expected to double in the next twenty years. It requires 600,000 gallons of water to produce one ton of synthetic rubber. The daily consumption of the average household in the United States is 400 gallons.

About three quarters of the water used in most humid industrial areas comes from surface waters (rivers, lakes, artificial reservoirs, etc.). The rest comes from groundwater. Pollution and waste still prevent maximum use of surface waters, on which we depend.

Over one fourth of the earth's land surface is desert, and dam construction can be vital in these areas. The Aswan Dam in Egypt brought almost 2.5 million acres of new land into cultivation and generates 10 billion kilowatt-hours of electricity each year. Desalinization of sea water, although used in some arid areas, is still too expensive for general use.

Water conservation is a pressing world need since supplies, although they are never exhausted but are replenished in the water cycle, can never be increased. They can, however, be more efficiently used and distributed. Pollution of water supplies by domestic and industrial wastes can upset the delicate ecological balance, and has very serious biologic, economic, and recreational effects.

Arid and semiarid regions cover about 1/3 of the earth's surface. This water hole in Pakistan is typical of local water supplies.

Lake Solitude, Wyoming, a typical area of recent glaciation

GLACIERS AND GLACIATION

We live today in the twilight of a great episode of refrigeration, when much of the Northern Hemisphere was covered by continental ice sheets, like those that still cover Antarctica and Greenland. Although the ice itself has now retreated from most of Europe, Asia, and North America, it has left traces of its influence across the whole face of the landscape in jagged mountain peaks, gouged-out upland valleys, swamps, changed river courses, and boulder-strewn, table-flat prairies in the lowlands.

Glaciers are thick masses of slow-moving ice. In the higher lands and polar regions, the annual winter snowfall usually exceeds the summer loss by melting. Permanent snow fields build up, and their lowest boundary is the snow line, the actual height of which varies with latitude and climate. Buried snow recrystallizes to form ice, which moves slowly under its own weight. It moves most rapidly in the middle of the glacier.

GLACIAL EROSION has a powerful effect upon land that has been buried by ice and has done much to shape the mountain ranges of our present world. Both valley and continental glaciers acquire many thousands of boulders and rock fragments, which, frozen into the sole of the glacier, gouge and rasp the rocks over which the glaciers pass. The rocks are slowly abraded down to a smooth, fluted, grooved surface. Glacial meltwater, from periods of daylight or summer thaw, seeps into rock fissures and joints. When it freezes again, it helps to shatter the rocks, some of which may become frozen into the body of the glacier and be carried away as the glacier moves downslope. Avalanches and undercutting of valley sides add to the rock debris.

ROCK SURFACES display fluting, striation, and polishing effects of glacial erosion. The form and direction of these grooves can be used to show the direction in which the ice moved.

TYPICAL LONGITUDINAL SECTION of a valley glacier shows its structure and its profile of bedrock. Surface ice is brittle, but underlying ice crystals bend, shear, and glide, causing ice to flow by deformation.

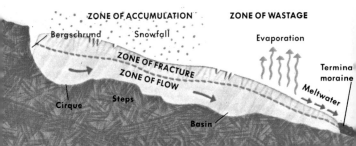

ZONE OF ACCUMULATION   ZONE OF WASTAGE

Bergschrund   Snowfall   Evaporation

ZONE OF FRACTURE

ZONE OF FLOW

Terminal moraine

Meltwater

Cirque   Steps

Basin

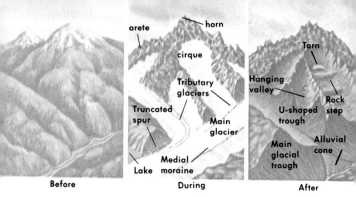

Before

During

After

Labels on diagram: arete, horn, cirque, Tributary glaciers, Truncated spur, Main glacier, Lake, Medial moraine, Tarn, Hanging valley, Rock step, U-shaped trough, Main glacial trough, Alluvial cone

THREE STAGES OF GLACIAL EROSION are illustrated above, showing mountain country before, during, and after glaciation. Glaciers cut U-shaped valleys, modifying and deepening the interlocking pattern of earlier meandering river erosion. The valleys are straightened and truncated by ice flow. Tributary hanging valleys develop where the rate of erosion by tributary glaciers is lower than that of the glacier in the main valley. As the ice retreats, waterfalls flow out of them into the main valley.

HORNS, ARETES, AND CIRQUES are all products of glacial erosion. Horns are sharp, pyramidal mountain peaks formed when headward erosion of several glaciers intersect. Aretes are sharp ridges formed by headward glacial erosion. Continental glaciers tend to produce a smoothed-out effect on the landscape, such as that of the Laurentian Shield in Canada. Cirques are bowl-shaped valleys formed at heads of glaciers and below aretes and horned mountains; often contain a small lake, called a tarn.

U-shaped glaciated valley, Clinton Canyon, New Zealand

Horns and aretes in glaciated area, Switzerland

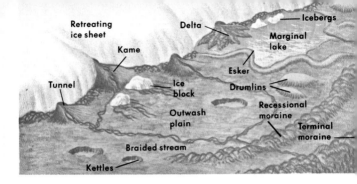

Labels in figure: Retreating ice sheet, Delta, Icebergs, Kame, Marginal lake, Esker, Tunnel, Ice block, Drumlins, Outwash plain, Recessional moraine, Terminal moraine, Braided stream, Kettles

IDEALIZED GLACIAL LANDSCAPES show typical depositional features. They are collectively called till deposits. Glacial deposits of rock fragments are carried by the glacier on its surface within the ice and at its base. This material is deposited either beneath or at the foot of the ice field, forming unsorted and unbedded "till." Meltwater streams flowing from the glacier form sorted, stratified glaciofluvial or outwash deposits. These and other glacial deposits are often described as "drift." Deposits also occur during retreat of ice.

MORAINES are deposits of glacial till formed either as arcuate mounds at the snout of the glacier (terminal moraines) or as sheets of till over considerable areas (boulder clay). Successive terminal moraines often mark retreat stages of glaciers (recessional moraines). Moraines are made up of a variety of unsorted rock fragments in unbedded clay matrix.

ERRATICS are boulders of "foreign" rock carried by glaciers. Some are up to 100 feet across, and most are found many miles

DRUMLINS are low, rounded hills, sometimes reaching a mile in length, found in glaciated areas. Aligned in the direction of ice flow, their steeper, blunter ends point toward the direction from which the ice came. They are formed by plastering of till around some resistant rock mass. They produce a characteristic "basket-of-eggs" topography.

from their points of origin. They often have blunted edges and rather smooth faces, but most lack glacial striations.

KAMES are isolated hills of stratified material formed from debris that fell into openings in retreating or stagnant ice. Kame terraces are benches of stratified material deposited between the edge of a valley glacier and the wall of the valley.

GLACIOFLUVIAL DEPOSITS are all sorted and bedded stream deposits. Outwash deposits, formed by meltwater streams, are flat, interlocking alluvial fans.

ESKERS are long, narrow, and often branching sinuous ridges of poorly sorted gravel and sand formed by deposition from former glacial streams.

KETTLE HOLES are depressions (sometimes filled by lakes) due to melting of large blocks of stagnant ice, found in any typical glacial deposit.

GLACIAL LAKE DEPOSITS are formed either by meltwater or the blocking of river courses. Deltas and beaches mark the levels of many such lakes.

VARVE CLAYS are lake deposits of fine-grained silt, showing regular seasonal alternations of light-colored, thicker bands deposited during wet, summer months and thinner, dark bands representing the finer, often organic, material of winter deposits that settle below the frozen lake surface.

Maximum extent of Pleistocene ice sheets and glaciers

ANCIENT PERIODS OF GLACIATION produced land features still in evidence today. The features already described can be seen in connection with existing glaciers, but older glacial deposits and erosional features prove the occurrence of earlier glacial episodes. The most recent of these is the Pleistocene glaciation which began about two million years ago. It involved four major episodes of glaciation, when continental ice sheets covered about one quarter of the earth's surface, including parts of North America, northern Europe, and northern Asia. Glacial advances were separated by warmer, interglacial periods. In the areas outside those covered by glaciers, especially in the Southern Hemisphere, corresponding pluvial periods of abnormally heavy rainfall marked Pleistocene times, probably caused by changes in the general pattern of wind circulation produced by continental glaciers.

Pleistocene glaciation molded such familiar features of our present landscapes as the jagged peaks of the Rockies and the Alps, the rich farm soils of the northern midwestern states, and the Great Lakes.

GLACIATION is important because it has molded the topography of much of the Northern Hemisphere. It also poses a number of basic geologic problems.

CHANGES IN SEA LEVEL of up to 300 feet occurred when much of the ocean's water was locked in continental glaciers. Even today, if present glaciers and ice sheets that cover 10 percent of the earth's surface were to melt, sea level would rise by some 300 feet. The continental margins would be flooded, and many of the world's major ports would be submerged. This may happen again. If it does not, and we are instead living in an interglacial rather than postglacial episode, then continental glaciers may again spread across much of the earth.

LOADING THE CRUST by ice caused sags to develop which reached about 1,500 ft. under the thickest ice. With the melting of the glaciers, the crust began to rise again, and the history of this rise can be traced in Scandinavia, North America, and Europe. The crust rises about 9 inches per century.

CAUSES OF ICE AGES are still unknown. Possible causes may be changes in the broad pattern of the circulation of the oceans, changes in the relative position of the earth and sun, changes in solar radiation, and the presence of some blanket such as volcanic dust to reduce solar radiation reaching the earth. It has also been suggested that the Pleistocene glaciation of the Northern Hemisphere could have been the result of surges in the Antarctic ice sheet producing wide ice shelves around the Southern Ocean, which cooled the Northern Hemisphere.

PRE-PLEISTOCENE GLACIATIONS are much less easy to detect than the very recent Pleistocene. One major episode of glaciation took place in the Southern Hemisphere in Permo-Carboniferous times, about 230 million years ago. Tillites (indurated glacial tills) and striated rock pavements show large areas of Australia, South America, India, and South Africa to have been glaciated. The character of these glacial deposits suggests that these areas, now remote, formed a single massive continent (Gondwanaland) at that time.

There is also evidence of an Ordovician (about 450 million years ago) and a late Pre-Cambrian glaciation (about 600 million years ago).

UPLIFT,
IN METERS
Since 6800 B.C.

50
100
150
200
200
150
100
50
0

Gulf of Bothnia

Gulf of Finland

THE OCEANS

Oceans play a major role in the earth's natural processes because of their production and control of climate, supplying moisture to the atmosphere and providing a vast climatic regulator. They form the ultimate site of deposition of almost all sediment and are the home of many living species of animals and plants. The oceans cover over 70 percent of the earth's surface. The continents are surrounded by shallow, gently sloping continental shelves. The average ocean depth is almost three miles, but trenches, up to 36,000 feet deep, are found in places. Although much of the deep ocean floor is a flat plain, some parts are more mountainous than the mountain regions of dry land. A worldwide system of midoceanic ridges includes submarine mountain chains, marked by intense vulcanism and earthquake activity, and offset by transform faults. They are the sites of the formation of new crustal rocks (p. 141). There are also many volcanic islands, including many submerged below sea level.

California Coast shows force of breaking waves eroding shoreline.

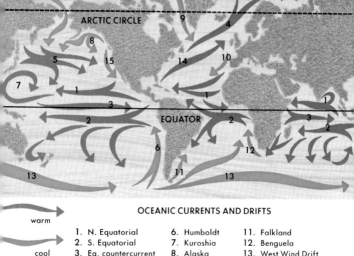

ARCTIC CIRCLE

EQUATOR

OCEANIC CURRENTS AND DRIFTS

warm

cool

1. N. Equatorial	6. Humboldt	11. Falkland
2. S. Equatorial	7. Kuroshia	12. Benguela
3. Eq. countercurrent	8. Alaska	13. West Wind Drift
4. N. Atlantic drift	9. Labrador	14. Florida
5. N. Pacific	10. Canaries	15. California

CURRENTS have a major influence on world weather patterns. Differences in the density of sea water of varying salinity and differences in temperature produce water circulation in the oceans. The colder, more saline, denser water sinks downward to produce deep ocean currents. Nearer the surface of the sea, the combined influence of winds and the rotation of the earth produce the more familiar surface currents, including the Gulf Stream. These surface currents follow great, swirling routes around the ocean basins and the equator. Some currents move at speeds of over 100 miles a day.

SEA WATER includes about 3.5 percent of dissolved chemicals by weight. Salt (NaC1) is the most common solute, with smaller quantities of magnesium chloride, magnesium and calcium sulfates, and traces of about 40 other elements. Salinity is the number of grams of these dissolved salts in 1,000 grams of sea water. Although the proportions of these salts to one another are very similar throughout the oceans of the world, the total salinity of the oceans varies from place to place and with depth. It is low near river mouths, for example, and high in areas of high evaporation.

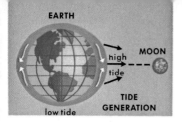

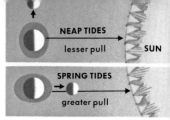

TIDES are twice-daily movements of billions of tons of ocean water influenced by many factors on the surface of the earth as well as from space. Mainly, the gravitational pull of the moon upon the earth causes the waters to bulge toward it twice a day, creating what we call high tides. The corresponding bulge or high tide on the distant side of the earth from the moon is caused by the corresponding lower attraction of the moon at this greater distance, allowing the oceans to "swing" outward. Tides rise only two or three feet on open coastlines, but in restricted channels can reach fifty feet.

WAVES are produced chiefly by the drag of winds on the surface of water. The water is driven into a circular motion, but only the wave form, not the water itself, moves across the ocean surface. Waves generally affect only the uppermost part of the oceans. Wave base is half the wave length of any particular wave system. When they run into shallow water, waves drag bottom, and the topmost water particles break against the shore. Waves play an important part in the shaping of coastlines, both in sediment transport and in erosion. Some large waves (tsunami) are caused by earthquakes.

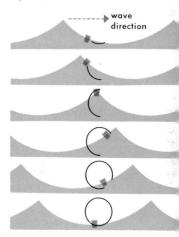

wave direction

Breaker
wave height increases

Beach

wave length decreases

Deep water waves not affected by bottom

NORTH AMERICA

Rift valley

Continental shelf

Mid-ocean Canyon

Continental slope

EUROPE

Continental shelf

Continental slope

MID-ATLANTIC RIDGE

SOHM ABYSSAL PLAIN

Islands

Transform faults

AFRICA

THE EDGES OF THE CONTINENTS are commonly marked by margins of broad, flat shelves which slope gently (at about 1:1000) to a depth of about 450 feet. At this depth, they merge into the steeper continental slope. The width of shelves varies from a few miles to 200 or more miles. Commonly, shelves are about 30 miles wide. The shelves seem to be formed by the deposition and erosion of fairly young sediments, many of them of Pleistocene age. Changes in sea level of some 500 feet have been involved during this period.

THE CONTINENTAL SHELF AND SLOPE rim the continents, the wide shelf dropping off steeply at the slope to the depths of the seafloor. At the base of the continental slope, there is often a convex *rise*, formed from slumped sediments. This area ranges from a few miles to about 100 miles in width.

The great vertical exaggeration of the diagram suggests a much steeper profile than really exists. Even so, the slope is almost a hundred times steeper than the shelf.

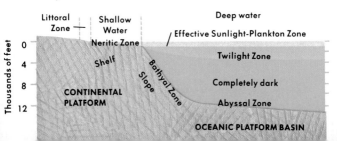

Littoral Zone

Shallow Water Neritic Zone

Deep water

Effective Sunlight-Plankton Zone

Shelf

Twilight Zone

Bathyal Zone

Slope

Completely dark

Abyssal Zone

CONTINENTAL PLATFORM

Thousands of feet

0

4

8

12

OCEANIC PLATFORM BASIN

SUBMARINE CANYONS cut through the continental shelves and slopes and are widely distributed along the edges of continents. Some seem to be continuations of rivers on the land, but others show no such relation to drainage and do not extend across the continental shelves. All canyons tend to have a V-shaped profile and to have tributary systems much like those of terrestrial rivers. Their deeper mouths are marked by great deltalike fans of sediment, which gradually build up to form the continental rise.

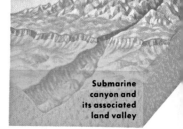

Submarine canyon and its associated land valley

Submarine canyons are thought to be formed by the erosion of turbidity currents, which sometimes attain considerable velocity. Heavy with silt, they have a strong scouring and erosive power.

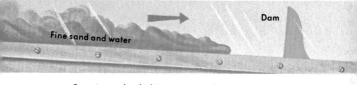

Fine sand and water

Dam

Experimental turbidity current in a laboratory tank

TURBIDITY CURRENTS are dense, flowing masses of sediment-carrying water flowing at speeds of up to 50 miles per hour. Many are probably triggered by earthquake disturbances of unconsolidated sediment on the continental shelves and slopes. The coincidence of some submarine canyons with river courses, such as those of the Hudson and Congo, has been thought to be the result of river erosion at earlier periods of lower sea level. Probably it is the result of either the presence of thicker, unstable masses of sediment near river mouths or turbidity flow from river mouths in times of flood.

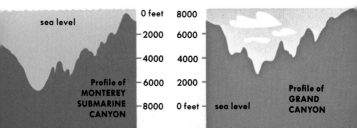

sea level

| 0 feet |
| 2000 |
| 4000 |
| 6000 |
| 8000 |

Profile of
MONTEREY
SUBMARINE
CANYON

8000
6000
4000
2000
0 feet — sea level

Profile of
GRAND
CANYON

Coastal view of Hargrave's Lookout, New South Wales, Australia

COASTLINES mark the boundaries of land and sea. Although the great variety of rock types, structures, currents, tides, climate, and fluctuating sea levels produce many different types of coastlines, each can be understood as the product of three simple processes: erosion, deposition, and changing sea level.

Coastal erosion is the result of the twice-daily pounding by the sea, wearing down the margins of the land, creating coastal features, and cutting back the shoreline at a rate of several feet a year.

CLIFFS AND WAVE-CUT PLATFORMS

are characteristic of shorelines undergoing erosion. Waves undercut the rocks near sea level.

CAVES

are formed by erosion along a conspicuous line of weakness in a cliff, such as along joints and faults. Continuing erosion may form an arch.

BAYS AND HEADLANDS

are produced by relative differences in the resistance to erosion of coastal rocks. The more resistant standout as headlands, but ultimately, the concentration of wave erosion on the headlands and deposition in the bays have a tendency to produce a straight coastline.

Drowned Valley, view from Mt. Wellington, Tasmania

A CHANGING SEA LEVEL is represented by many features around coastlines. Within historic times, established towns have been submerged. Raised beaches and wave-cut platforms are common in many areas, rising many feet above present sea level. Far inland and high on the slopes of mountains, fossils of marine animals give further proof of older and more profound changes in sea level, reflecting major changes in the geography of the past. Submergence and emergence of coastlines modify the general features of erosion and deposition.

EMERGENT COASTLINES are less common than the submerged. They are marked by raised beaches and cliffs, and often by an almost flat coastal plain, sloping gently seaward.

SUBMERGED COASTLINES, due to postglacial rising sea level, are indented coastlines with deep inlets and submerged glacial valleys. The "grain" of the coastline depends upon the character and structure of the rocks. In Atlantic-type coastlines, the structural trends are more or less perpendicular to the coast. In Pacific type, they are parallel to the coast.

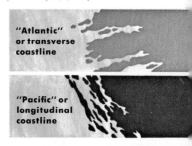

"Atlantic" or transverse coastline

"Pacific" or longitudinal coastline

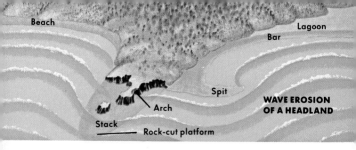

Beach

Lagoon

Bar

Spit

Arch

WAVE EROSION OF A HEADLAND

Stack

Rock-cut platform

MARINE DEPOSITION may be recognized as the dominant process where shorelines are marked by a number of familiar features. Ultimately, the balance of coastal erosion and deposition tends to produce a coastline in temporary equilibrium. Although this is more quickly formed in soft strata, it requires thousands of years in resistant rocks.

BEACHES consist of sediment sorted and transported by waves and currents. Most of the sediment is the size of sand or of gravel, but local cobble and boulder deposits are also common. Beaches vary greatly, depending upon the sediment supply, the form of the coastline, wave and current conditions, and seasonal changes. On irregular coastlines, they tend to be confined to the bays. Oblique waves and longshore currents often produce constant lateral movement of beach material. Beaches thus exist in a state of dynamic equilibrium, as a moving body of wave-washed and sorted sediment.

Interaction of river and marine deposition and erosion, Wales

OFFSHORE BARS formed of sand and pebbles are generally separated from the main shoreline by narrow lagoons. They are ancient beach deposits that are characteristic of submerged coasts. In North America, offshore bars are common along the Atlantic and Gulf coasts. They are generally parallel to existing coastlines.

SANDBARS are formed on indented coastlines by longshore drift of sediments parallel with the coast. When sediment is carried into deeper water, as in a bay, the energy of the waves or currents is reduced, and the sediment is deposited as a bar. The slower the current, the more rapidly the sediment is deposited. The bar is an elongation of the adjacent beach, partly or completely cutting off the bay. Spits are bars that extend into open water rather than into a bay. The free ends of many spits are curved landward by wave refraction. Spit growth may lead to blocking of harbors and diversion of rivers that flow into the sea, sometimes requiring dredging.

Typical cliff and beach scenery shows balance of erosion and deposition. (Dorset, England)

DELTAS (p. 45) are formed by rapid deposition of material carried by a river when it enters the deep water of a lake or the sea and loses its velocity. All deltas have a similar pattern of deposition and sediment distribution. Shape and size vary, depending upon local conditions.

STORM BEACHES (or berms) are formed by storms that throw gravel and boulders up above the normal high-tide level.

Wide, sandy beach with storm beach near cliffs at Lavernock, Wales

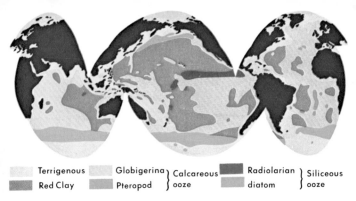

| | Terrigenous | | Globigerina | } Calcareous | | Radiolarian | } Siliceous |
| | Red Clay | | Pteropod | } ooze | | diatom | } ooze |

MARINE SEDIMENTS are classified in three broad categories: littoral, neritic, and deep-sea sedimentation. Littoral sediments form between high- and low-tide levels; neritic sediments accumulate on the continental shelf. These two groups of shallow-water sediments cover only about 8 percent of the ocean floor. They are made up of a variable mixture of terrigenous or land-derived debris, chemical precipitates, and organic deposits. They differ from place to place due to variations in coastlines, rivers, and changes in sea level.

In general, in shallow-water sediments, there is a direct relationship between the size of a sediment particle and the distance to which a given current will carry it, the larger particles being deposited nearer the source. This pattern is modified by the action of waves, currents, and turbidity currents.

Deep-sea sediments deposited outside the continental shelf form a layer generally less than 2,000 feet thick over the deeper parts of the ocean floor. They are much thinner and younger in age than we should predict from knowledge of present rates of sedimentation. The abyssal parts of the ocean show much more uniform sediments than the continental margins, where most terrestrial debris is deposited. Those of the bathyal zone, deposited on the continental slopes to a depth of about 12,000 feet, include muds of various kinds. The sediments of greater abyssal depths are red clays and various oozes which cover 30 percent and 47 percent of the ocean floor respectively.

CALCAREOUS OOZES are fine-medium-grained sediments that cover almost half the ocean floors. They occur down to a depth of almost 15,000 feet. Below that depth, the calcareous tests of the planktonic foraminifer *Globigerina* and pteropod molluscs, which form most of the sediments, are dissolved.

SILICEOUS OOZES are derived from the remains of surface-living organisms (*diatoms* and *radiolaria*) forming very slowly at great depth in the oceans. Diatom oozes are found chiefly in polar seas where predatory creatures are less common. Radiolarian ooze is most commonly found in the warm, tropical waters.

DARK RED CLAY is formed from meteoric dust and from very fine terrigenous or volcanic particles carried by the wind or in suspension in sea water. It may form as slowly as one inch every 250,000 years. It may include volcanic ash layers. Other sediments, such as manganese nodules (shown here), are present in some parts of the ocean floor.

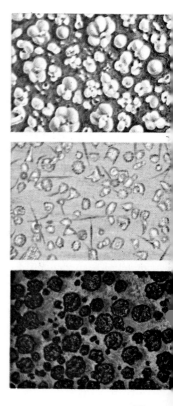

WIND is most effective in transport and deposition in deserts, near shorelines, and in other places where there is a supply of dry, fine-grained, loose sediment with little vegetation to hold it together. Great Sand Dunes National Monument, shown here, is formed by deposition of windborne sand against mountain range.

WINDS

Winds are movements of the atmosphere brought about not only by the rotation of the earth but by unequal temperatures on the earth. The heat of the sun, the chief source of this circulation, is more concentrated in the tropics than in high latitudes. This produces vast atmospheric convection currents, having a broad, constant overall distribution that reflects the earth's rotation. It also shows wide local variations in speed and direction due to differences in topography and other atmospheric conditions.

A major role is played by the wind in the distribution of water from the oceans to the land. Water vapor, in turn, has a blanketing effect that keeps the earth's surface temperature higher than it would otherwise be. Wind is an agent of transport and, to a lesser extent, of erosion. In this, it resembles flowing water, but because of its much lower density (only about 1/800 that of water), it is far less effective and generally transports only the finer dust particles. Dust from volcanic explosions will often give brilliant sunsets in distant lands for many months after the explosion. Winds transport through the atmosphere comparatively large quantities of salt crystals gathered from the ocean's surface.

WIND EROSION is very limited in extent and effect. It is largely confined to desert areas, but even there it is limited to a height of about 18 inches above ground level.

DESERT PLATFORMS are clean, windswept areas where pebbles may have been rolled and bounced along by the force of strong winds. Larger cobbles and boulders are left behind.

VENTIFACTS, found in deserts, are pebbles or cobbles that have developed polished surfaces and sharp edges under wind abrasion.

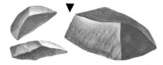

DESERT EROSION tends to expose bare rock surfaces, which may stand up without a cover of vegetation, as in Pakistan.

Semiarid landscape has distinctive erosional character.

WIND DEPOSITS consist of transported particles that are effectively sorted according to size because of the limited carrying capacity of the wind. Sand dunes, for example, generally consist of sand grains of more or less uniform size, which are rounded and pitted or frosted by abrasion.

Sand dunes are found in areas where there is a large supply of dry, loose, fine-grained material. Like snow-drifts, they form around local obstructions. They also migrate downwind. Their particular size and form depend upon the sand supply, the presence of vegetation, and the velocity and constancy of direction of the prevailing wind. Dunes may be transverse or longitudinal to the wind direction.

BARCHANS are crescentic dunes that often build up to 400 yards long and 100 feet high, and are formed mostly in deserts with more or less constant wind directions. They are not static and may migrate up to 60 feet per year. The crescent points show downwind direction. Winds also produce giant ripple marks on sand surfaces. Barchans usually are found in groups, or swarms, and may form long lines, or chains, across a plain.

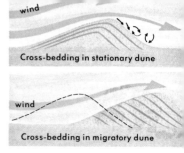

Cross-bedding in stationary dune

Cross-bedding in migratory dune

SAND-DUNE SHAPE typically has gentle windward slope and steep leeward slope, down which sand grains slide or roll. Dotted line shows how continuous movement of sand grains produces migration of whole sand dunes. Dunes take many forms. It would not be easy to compile a complete list of all the varieties. Variations in form include scalloped sides and irregularities in plan of the crest.

ANCIENT DUNE DEPOSITS, preserved in such sedimentary rocks as those of the Navajo Sandstone in Zion Canyon National Park, display aeolian bedding, sorting, and sand grain rounding similar to those of present-day dunes. Careful mapping of bedding directions reveals ancient wind directions. In this way, it has been possible to make a map of the Permian winds of southwestern United States 225 million years ago.

Cross-bedding, in wind-deposited sandstone, reflects its formation in ancient sand dunes, east of Echo Cliffs, Arizona.

LOESS DEPOSITS, formed of fine-grained silt, lack any bedding but often have vertical joints. Transported by wind from deserts, from dried-up flood plains, from river courses, or from glacial deposits, they are common in midwestern United States, China, Europe, and in many areas surrounding the world's deserts and glacial outwash areas. Loess produces fertile soils, partly beause of its very high porosity. Loess is yellow or buff in color and often forms vertical cliffs. Artificial caves in easily worked loess may provide homes.

PRODUCTS OF DEPOSITION

Sedimentary rocks are generally formed from the breakdown of older rocks by weathering and the agents of erosion described on pp. 34-76. A few are chemical precipitates, or organic debris. Sedimentary rocks cover about 75 percent of the earth's surface.

CLASTIC OR DETRITAL SEDIMENTARY ROCKS are formed from the debris of preexisting rocks or organisms. The weathered rock material is generally transported before it is deposited. This movement often gives round grains. The debris is eventually laid down in horizontal layers, usually as marine deposits but sometimes as deposits from rivers, lakes, glaciers, or wind. Clastic rocks are solidified sediments.

CONGLOMERATE consists of rounded pebbles or boulders held tightly in a finer-grained matrix. The pebbles are usually of quartz at least 1/4 inch or more in diameter.

SANDSTONE consists of sand-size particles, usually of quartz. It may show considerable variation in cementing minerals, in rounding and sorting of particles, and in forms of bedding.

ARKOSE is a quartz-feldspar sandstone usually formed in desert areas by rapid erosion and deposition of feldspar-rich igneous rocks.

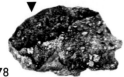

CALCARENITE consists of broken shells or other organic material and fragments from older limestones. It is deposited as sedimentary debris.

GRAYWACKE is a poorly sorted mixture of rock fragments, quartz, and feldspar fragments in a clay matrix. Often formed by turbidite flows (p. 67), graywacke always indicates rapid erosion and deposition under unstable conditions.

SHALE consists of very fine-grained particles of quartz and clay minerals. It is consolidated mud that has been deposited in lakes, seas, and similar environments. About 45 percent of all exposed sedimentary rocks are shales.

ORGANIC SEDIMENTARY ROCKS are formed from organic debris—the deposits or remains of once-living organisms (shells, corals, calcareous algae, wood, plants, bones, etc.). Although they are a form of clastic rock, organic rocks may contain more and better-preserved fossils, as they are laid down near the place where the animal or plant once lived.

CHEMICALLY FORMED SEDIMENTARY ROCKS consist of interlocking crystals precipitated from solution. They, therefore, lack the debris-cement composition of other sedimentary rocks. A decrease in pressure, an increase in temperature, or contact with new materials may cause minerals to precipitate from solution.

LIMESTONE consists chiefly of calcite from concentrated shell, coral, algae, and other debris. It may grade into dolomite, characterized by the presence of calcium and magnesium carbonate (p. 28). Chalk is a fine-grained limestone of minute coccoliths. Travertine is limestone precipitated by springs.

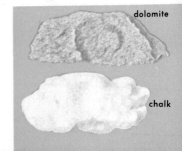

dolomite

chalk

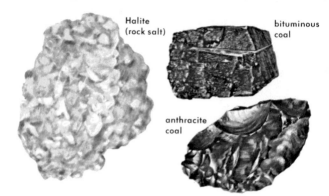

Halite
(rock salt)

bituminous
coal

anthracite
coal

EVAPORITES are chemical precipitates, formed by evaporation in shallow, land-locked basins of water. They vary greatly in texture and composition. Rock salt, gypsum, anhydrite, and potassium salts are the most common. Important industrial minerals.

COAL is consolidated peat, formed by the decomposition of woody plant debris (p. 98). It is an organic rock, and plant structures may still be preserved in it. Coals grade from lignite to anthracite, which has about 95 percent carbon in it.

SEDIMENTARY ROCKS are important natural resources. Shales and limestones are used for cement, clays for ceramics; other rocks are used for road metal. Sedimentary iron ores, bauxite, and coal form the basis of much heavy industry; soil is the ultimate basis of most of our food supplies.

A typical coal mine operation

THE SIZE, SHAPE, AND SORTING of sedimentary structures within rocks may provide clues to the depositional environment that existed during their formation. Rounded, frosted, well-sorted grains, for example, indicate wind-deposited sand.

CROSS-BEDDING is a term applied to sweeping, arclike beds that lie at an acute angle to the general horizontal stratification. It is common in stream and deltaic deposits, in deeper marine waters, and in sand dunes. Cross-bedding reflects the direction of current flow.

GRADED BEDDING is due to differential settling of mineral grains, and is useful for determining the correct "way up" in folded strata. The layers of coarse rock have a sharp base and gradually grade upward into finer-grained materials.

VARVED BEDDING is a type of thin, graded bedding that has alternate laminations of coarse and fine-grained material. Especially common in glacial lake deposits, it is characteristic of seasonal deposition, each pair of varves representing a year. (See page 60.)

MUD CRACKS form where lake and mud deposits are dried by the sun and are preserved by burial under mud.

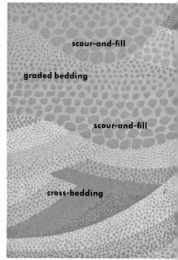

scour-and-fill

graded bedding

scour-and-fill

cross-bedding

SCOUR-AND-FILL structures are formed by erosion and subsequent filling of valleys and channels in a river bed.

RIPPLE MARKS are produced by wind in desert sand deposits and by waves or currents in various aqueous environments.

THE CRUST: SUBSURFACE CHANGES

Running water, ice, wind, and other agents of erosion slowly wear down the surface of the continents. Over the earth as a whole, some 8 billion tons of sediment is carried by rivers into the sea every year, equivalent to some 200 tons per square mile of land surface. This represents an average lowering of the rivers' drainage areas by approximately one foot every 9,000 years.

The ultimate effect of this erosion would be to reduce the continents to a flat surface, but there are two earth processes that tend to interrupt this continuing erosion and restore the balance: the tectonic uplift of areas of both the existing continental and offshore areas of sedimentation (diastrophism), and the broadly related process of igneous activity (p. 83). Most of these forces act slowly, over long periods of time, and the earth's crust is always in a state of dynamic equilibrium, though continuously changing, as a result of their varying interaction.

THE ROCK CYCLE

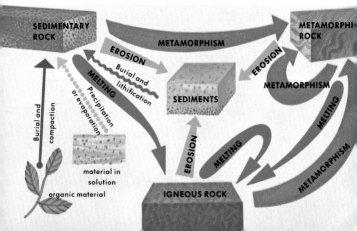

Volcanic cones stand out as hills near Springville, Arizona.

Igneous rocks form the foundations of the continents, but most of the surface of the continents is made up of sedimentary rocks of various ages. These layers have been deposited on and around ancient continental cores or shields, which are made chiefly of granitic igneous and metamorphic rocks. The cores of mountain chains generally reveal the same rocks. These igneous rocks were generally formed at great depths and later uplifted, eroded, and covered with a relatively thin veneer of sediments.

VOLCANOES

Volcanoes are mountains or hills, ranging from small conical hills to peaks with 14,000-foot relief, formed by lava and rock debris ejected from within the earth's crust. Of the more than 500 active volcanoes, some extrude molten lava, others erupt ash and solid (pyroclastic) fragments. Still others eject both, and all emit large quantities of steam and various gases. Some eruptions are relatively "quiet," others explosive.

VOLCANIC ACTION results in the formation of five basic types of volcanoes, but no two are ever quite alike. The kinds of materials that erupt from a volcano largely determine the shape of its cone. Fluid streams of lava travel far and usually produce wide-based mountains; ash, viscous lavas, and cinders usually build up steep cones.

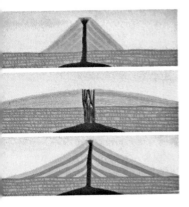

CINDER CONES are steep-sided, symmetrical cones, such as Vesuvius, formed by the eruption of cinders, ash, and other pyroclastic products.

SHIELD VOLCANOES, like those in Hawaii, are broad domes formed by lava flows from a central vent or from fissures and parasitic vents.

COMPOSITE CONES are formed from interbedded lava flows and pyroclastic debris. They are intermediate in form between cinder and shield volcanoes.

CALDERAS are formed by the collapse of the top of a volcano following an explosion. New cones may be born in the caldera. Crater Lake in Oregon is an example of this type of volcano.

PLATEAU BASALTS cover great areas in the Columbia River Valley, Iceland, and India, and are apparently extruded, to spread in thin sheets, not from central vents but from cracks or fissures.

VOLCANIC PRODUCTS include bombs, cinders, ash, and dust, as well as lava and gases. Most lava has a basaltic composition (p. 93) consisting of plagioclase feldspars, pyroxene, and olivine. Erupted at temperatures of between 900° and 1,200°C, it can flow for great distances at considerable speeds.

THE TYPES OF LAVA differ in form and texture. Pillow lava (left) is formed by submarine volcanoes. Blocky lava has a jagged surface and is produced by sudden gas escape. Ropy lava (right) forms at a higher temperature than blocky lava.

VOLCANIC BOMB shows spindle-shaped form.

VOLCANIC TUFF is a fine-grained, pyroclastic rock.

DISTRIBUTION OF VOLCANOES in narrow belts reflects earth's major plates (pp. 144-147) and is related to deep interior processes. Many lie around the Pacific Ocean. Others are formed in such areas of recent tectonic activity as the Mediterranean or the African Rift Valleys. Most oceanic islands are volcanic.

WORLD
DISTRIBUTION
OF
VOLCANIC
AREAS

INTRUSIVE IGNEOUS ROCKS are formed from magma rising within the earth's crust. Unlike the extrusive volcanic rocks, intrusive rocks crystallize below the earth's surface, and their presence becomes obvious only after the country rock into which they were intruded has been removed by erosion.

Intrusions show great variation in form. Some cut across bedding planes (discordant), while others run parallel with them (concordant). They range in size from dikes measuring a few inches wide to batholiths hundreds of miles across.

Intrusive rocks are often associated with important metallic mineral deposits, such as copper or nickel.

FORMS OF IGNEOUS INTRUSIONS

PLUTONIC, INTRUSIVE ROCKS, similar to granite, are identified by having a coarser crystal texture, produced by slower cooling than that in extrusive rocks. In general, rocks formed at shallow depths in the crust have an intermediate texture.

DIKES are usually vertical intrusive sheets, often discordant. They may occur in vast dike swarms, associated with a central volcanic neck or intrusive centers. Dikes and sills frequently have fine-grained, chilled contact margins.

Dikes, such as those on left, show chilled margins of fine-grained texture, where in contact with the country rock, as on right.

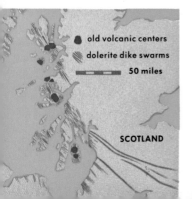

old volcanic centers
dolerite dike swarms
50 miles

SCOTLAND

Old volcanic neck at Shiprock, New Mexico, is circular in outline and fed a radiating series of once molten dikes.

VOLCANIC PLUGS, or necks, are more or less cylindrical, vertical-walled intrusions, generally of porphyritic rock. Whether or not they pass upward into lavas, or breccias, depends to a great extent upon the depth of local weathering.

SILLS are horizontal intrusive sheets, either concordant or discordant. Such sills as the 900-foot-thick Palisades on the Hudson River show vertical differentiation because of gravitational crystallization. Some sills and dikes are only a few feet in thickness; some develop columnar jointing similar to that of lava flows. Sills (A) are distinguished from lava flows (B) by the baking of overlying adjacent country rock by sills and erosional contacts in buried lava flows.

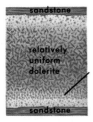

sandstone — fine-grained chilled zone 1% olivine

relatively uniform dolerite — **PALISADES SILL** — 25% olivine

sandstone — fine-grained chilled zone 1% olivine

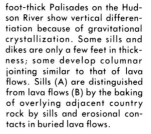

vesicular basalt — **LAVA FLOW (B)** — columnar joints

compact basalt — **SILL (A)** — columnar joints

LACCOLITHS are dome-shaped intrusions having a flat base but an arched concordant roof, formed by intrusive pressure, and a rather flat floor. They may show differentiation.

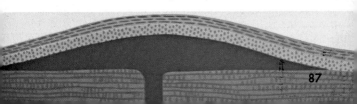

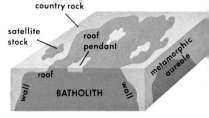

country rock

satellite stock

roof pendant

roof

metamorphic aureole

wall   BATHOLITH   wall

BATHOLITHS, or plutons, are major, complex, intrusive masses, generally granitic and often several hundred miles in extent. They are the largest intrusions and are found in areas of major tectonic deformation, as in the Idaho Batholith and in the continental shields. Although many plutonic granites show sharp contacts with older rocks, and are, therefore, intrusive in a strict sense, others show complete transition with surrounding country rocks and seem to result from the metamorphism or granitization of older sedimentary and metamorphic structures.

Stock and associated laccolith, Henry Mountains, Utah

stock

STOCKS are discordant, intrusive masses, a few miles in diameter. Some are plutonic, but others pass upward into ancient volcanic plugs. They are smaller than batholiths.

LOPOLITHS AND LAYERED INTRUSIONS are saucerlike intrusions, up to 200 miles across, with complex geologic histories. Distinct mineralogical layering is present, the more basic minerals generally being in the lower layers, as in the Bushveld complex in South Africa and the Sudbury, Ontario, lopolith. Gravity settling and convection currents seem responsible for this layering.

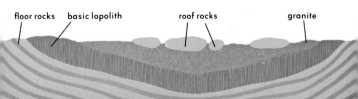

floor rocks   basic lopolith   roof rocks   granite

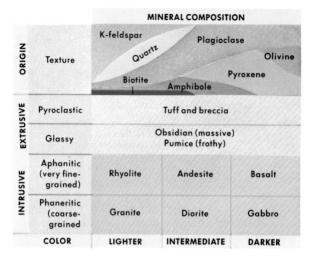

	MINERAL COMPOSITION		
ORIGIN — Texture	K-feldspar, Quartz, Biotite	Plagioclase, Amphibole	Olivine, Pyroxene
EXTRUSIVE — Pyroclastic	Tuff and breccia		
EXTRUSIVE — Glassy	Obsidian (massive) Pumice (frothy)		
INTRUSIVE — Aphanitic (very fine-grained)	Rhyolite	Andesite	Basalt
INTRUSIVE — Phaneritic (coarse-grained	Granite	Diorite	Gabbro
COLOR	LIGHTER	INTERMEDIATE	DARKER

CLASSIFICATION OF IGNEOUS ROCKS

We have already seen that igneous rocks vary in their occurrence, which tends to produce differences in texture (crystal size, shape, and arrangement). They also vary greatly in mineral content, and thus in chemical composition. Acid rocks (those containing quartz) make up the bulk of plutonic intrusions but seem to be confined to the continents, whereas basaltic rocks account for most of the volcanic rock of both the continents and the oceans.

Igneous rocks are commonly classified by their texture (the size, shape, and variation in their crystalline form) and their chemical composition (represented by their constituent minerals). These two factors reflect their rate of cooling and original magma composition.

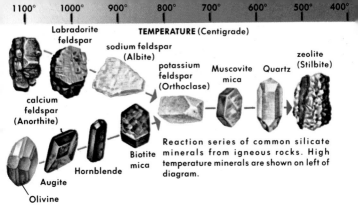

Labradorite feldspar

sodium feldspar (Albite)

potassium feldspar (Orthoclase)

Muscovite mica

Quartz

zeolite (Stilbite)

calcium feldspar (Anorthite)

Olivine

Augite

Hornblende

Biotite mica

Reaction series of common silicate minerals from igneous rocks. High temperature minerals are shown on left of diagram.

MAGMA is the molten silicate source material from which igneous rocks are derived. Although lava provides a surface sample of magma, the increased pressures and temperatures of deeper magmas permit a higher gas and water content than at the surface.

Igneous rocks show great variation in chemical composition, but this does not mean that each of the many types has crystallized from a different kind of magma. It seems probable that a single kind of basaltic magma is the parent of all varieties of igneous rocks, and that different chemical compositions result from crystallization differentiation.

Field and laboratory studies of igneous rocks show that igneous minerals have a definite sequence of crystallization: Iron, magnesium, and calc-silicate minerals (such as olivine, pyroxene, and calcium-plagioclase feldspars) form before sodium and potassium feldspars and quartz. This sequence is seen in differentiated intrusions.

Although crystallization of a basaltic magma would normally give a basalt (if the early-formed minerals are separated from the bulk of the magma by gravity settling or tectonic pressure), the remaining magma would be acid and relatively rich in silica and potassium and in sodium aluminosilicates. Continued crystallization and separation would then produce a rhyolitic magma. About 90 percent of the original magma would remain as crystalline rocks of basic composition.

The hypothesis of a single parent basaltic magma explains many otherwise puzzling features of igneous rocks, including the preponderance of basalt lavas and the frequent small and late rhyolitic flows in basaltic volcanic eruptions. Undisturbed cooling would produce granites, which would overlie basic rocks.

Other facts, however, suggest that such an explanation cannot account for all granitic rocks. The abundance of granites in mountain ranges, and their apparent continuity with metamorphic rocks, implies that many granites are formed by the "granitization" of deeply buried sedimentary rocks in the roots of mountain chains. This metamorphic origin of granitic, plutonic rocks might then provide "intrusive magmas" at higher levels in the crust. It seems unlikely that such huge masses of granite could have formed by differentiation of basic lavas.

Evolution of granitic magma from basaltic magma

A. **BASALTIC MAGMA**
50% SiO_2
10% $FeO + MgO$
40% other

B. Olivine, plagioclase feldspar, and pyroxene crystals form and settle

C. Iron and magnesium subtracted

Rich in FeO and MgO

D. **GRANITIC MAGMA**
70% SiO_2
2% $FeO + MgO$
28% other   melt

solids

Granite landscape in Sierra Nevada shows sheetlike weathering.

THE FORM AND TEXTURE of igneous rocks differ greatly in details of appearance and in the gross forms of the entire rock bodies. In contrast to volcanic lavas, which are extrusive igneous rocks, those formed at great depth are known as intrusive or plutonic (p. 89). These tend to crystallize more slowly and, therefore, have larger crystals than extrusive rocks. The texture of an igneous rock depends upon its rate of cooling, and thus on its geologic mode of formation (p. 89). The chemical and mineral content of igneous rocks depends upon the composition of the magmas from which they were formed. Magmas (or melts) rich in silica produce granitic (acid) type rocks. Those rich in iron and magnesium tend to be more mobile, and these produce basaltic (basic) type rocks. Because of their durability, many igneous rocks are used as road material. Large, specially cut blocks are used as ornamental building stones.

The color of an igneous rock is related to its composition. Acid rocks generally tend to be lighter in color than basic ones.

COMMON INTRUSIVE IGNEOUS ROCKS

GRANITE, usually light-colored and coarse-grained, contains about 30 percent quartz and 60 percent potash feldspar. It may be pinkish red or black-spotted. Granite is common in many large intrusions and often associated with mineral deposits.

GRANITE PORPHYRY has ground mass with longer crystals (phenocrysts) of feldspar, quartz, or mica. Very coarse-grained acidic rocks (pegmatites) have crystals over 40 feet long. Small porphyritic crystals are also found in lavas.

GABBRO is dark-colored and has a coarse, granitic-type texture consisting of plagioclase feldspar and pyroxene with traces of other minerals, but it contains no quartz.

COMMON EXTRUSIVE IGNEOUS ROCKS

RHYOLITE is a light, fine-grained volcanic rock of granitic composition, often porphyritic, with phenocrysts of quartz and orthoclase.

OBSIDIAN AND PUMICE have similar composition to rhyolite. Both are rapidly cooled. Obsidian is a translucent glass. Pumice is a rhyolitic froth characterized by cavities left by the release of gas.

BASALT, the most common lava, is dark, fine-grained, and rather heavy. It consists of pyroxene and a plagioclase feldspar. Cindery basalt (scoria) may have secondary minerals within the cavities formed by gas.

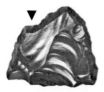

METAMORPHISM

Metamorphism is the process of change that rocks within the earth undergo when exposed to increasing temperatures and pressures at which their mineral components are no longer stable. Metamorphism may be *local*—contact metamorphism is due to igneous intrusion (pp. 86-99)— or *regional*—as takes place in mountain building, when slate, schist, and gneiss are formed. Metamorphism may take place in a solid state, without melting.

Effects of metamorphism depend upon the composition, texture, and strength of the original rock, and on the temperature, pressure, and amount of water under which metamorphism takes place. Metamorphosed rocks may differ in texture, mineral content, and total chemical composition from the parent rock.

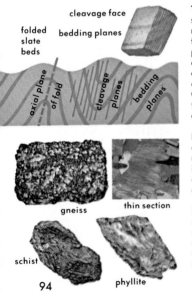

folded slate beds

cleavage face

bedding planes

axial plane of fold

cleavage planes

bedding planes

gneiss

thin section

schist

phyllite

TEXTURAL CHANGES are shown by most metamorphic rocks. Cleavage in slates (their fissility along definite planes) is produced by parallel realignment of such flaky minerals as mica, and is often inclined sharply to original bedding in the rocks. Cleavage is mainly the result of the increased pressure of dynamic metamorphism.

FOLIATION is the development of wavy or contorted layers under more intense metamorphism. It involves structural and mineralogical changes. Schists have closely spaced foliation.

PHYLLITES have a rather glossy, slaty appearance caused by recrystallization of flaky minerals along cleavage planes. Characteristics are in between schists and slates.

GROWTH OF NEW MINERALS results from intensive metamorphism, such as regional metamorphism involved in mountain building. Loss of some chemical components and addition of others may produce changes in total chemical composition of the original rock. Such flaky minerals as mica are unstable under these conditions and high-grade metamorphic rocks often have a granular appearance, developing whole new suites of metamorphic minerals. The particular assemblage depends upon the composition of the parent rock and the metamorphic environment. This has led to a concept of metamorphic facies.

MARBLE, fine to coarsely granular, is composed chiefly of calcite or dolomite. It derives from metamorphosed limestone.

QUARTZITE is a tough rock of metamorphosed quartz sandstone with a sugary texture. Its color is white to pink-brown.

GARNETIFEROUS SCHIST shows foliation in parallel arrangement of platy micas, and growth of garnet as a new mineral.

ECLOGITES contain garnet and pyroxene, formed from basic rocks at high pressure.

metamorphosed sedimentary rock

aureole, zone of contact metamorphism

STOCK (granitic intrusion)

RECRYSTALLIZATION of some minerals and the conversion of others is a feature common to many metamorphic rocks, especially those formed at high temperatures, such as those around intrusions (thermal metamorphism). Rocks around intrusions often display this as a result of contact metamorphism.

Black Marble

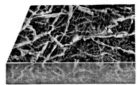

Quartzite

Eclogite

Garnetiferous schist

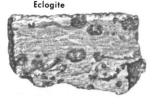

Offshore drilling rig symbolizes man's search for petroleum.

MINERALS AND CIVILIZATION

The foundation of 20th-century civilization is industrial, and the basis for industry is fuels and metal ores. The things that form the basis of life in the developed world—clean water supply, buildings, highways, automobiles, hardware, tools, fertilizers, plastics, fuels, chemicals, and more—ultimately come from the crust of the tiny planet on which we live.

Economic minerals are very unevenly distributed. Although most minerals themselves are widely scattered, deposits sufficiently rich to mine depend upon rare combinations of geologic processes. They are often found in isolated areas, and long geologic study may be needed to locate and exploit them. Almost 90 percent of the world's nickel supply, for example, comes from a single intrusion near Sudbury, Ontario.

New industrial processes and inventions bring demands for new minerals and fuels. The need for radioactive minerals spurred development of new techniques and dimen-

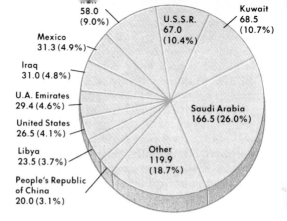

Iran
58.0
(9.0%)

Kuwait
68.5
(10.7%)

U.S.S.R.
67.0
(10.4%)

Mexico
31.3 (4.9%)

Iraq
31.0 (4.8%)

U.A. Emirates
29.4 (4.6%)

United States
26.5 (4.1%)

Libya
23.5 (3.7%)

People's Republic
of China
20.0 (3.1%)

Saudi Arabia
166.5 (26.0%)

Other
119.9
(18.7%)

WORLD CRUDE OIL proved reserves total about 640 billion barrels; distribution is concentrated in Middle East. (After U.S. Dept. of Energy)

sions in geologic surveys after World War II. The discovery of new mineral deposits can rapidly revolutionize the civilization of whole nations, as it has in the Middle East, where some of the world's largest petroleum resources have brought great wealth to Arab countries.

All mineral deposits are exhaustible. Once we have mined a vein of silver or a bed of coal, there is no way of replenishing it. This demands careful conservation of mineral deposits as well as long-term exploration and planning for new supplies.

Depletion of some essential minerals now poses serious long-term problems. It is estimated by some that 80 percent of the world's economically recoverable supplies of petroleum will be exhausted within a century. Some metals, including lead and copper, have comparably limited reserves. These estimates are based upon present rates of consumption, but an exploding world population could quadruple our mineral needs .

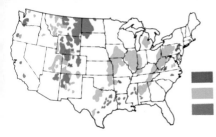

COAL FIELDS OF THE UNITED STATES

■ Anthracite

■ Bituminous

■ Sub-bituminous and lignite

MINERAL FUELS are basic to an industrial economy not only for heating, lighting, and transport but also for the industrial power needed in mineral processing, mining, and in manufacturing. Hydroelectricity, though of great importance in some areas, provides only a small fraction (less than 2 percent) of the world's power. Mineral fuels (oil, coal, and gas) provide about 98 percent. Coal and petroleum are fossil fuels.

Bituminous coal

Carnotite, a uranium mineral

COAL is a sedimentary rock formed from the remains of fossil plants. Buried peat, under pressure, loses water and volatiles and achieves a relatively high carbon content. Peat has about 80 percent moisture; lignite (an intermediate step between peat and coal), about 40 percent; bituminous coal, only about 5 percent. Anthracite, formed under conditions of extreme pressure, contains 95 percent carbon compared with bituminous coal, which has only about 80 percent. Most of the world's great coal deposits are in rocks of Pennsylvanian or Permian age.

ATOMIC FUELS at present supply only a fraction of the world's resources, but they will become more important. Mineral fuels used are ores of uranium.

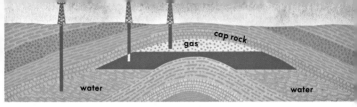

Cross section of oil field shows subsurface structure; petroleum is trapped in crest of an anticline sealed by impervious cap rock.

PETROLEUM is a general term for a mixture of gaseous, liquid, and solid hydrocarbons. When burned, this fossil fuel releases solar energy stored millions of years ago. Petroleum migrates from the source rock, where it forms, to other rocks. Most petroleum remains dispersed in rock pores and much escapes at the surface of the earth. But in commercial fields, it is trapped between an impermeable cap rock and a permeable reservoir rock, often floating on water, as illustrated.

OIL SHALE is carbonaceous shale that yields hydrocarbons when distilled. At present, the cost of oil and gas production from it is too high to make oil shale economically important, but it may be used in the future.

PETROLEUM EXPLORATION involves both geological and geophysical studies. Since most of the more obvious surface traps have now been discovered, seismic and other surveys are increasingly used to discover subsurface and submarine structures, such as those beneath the North Sea.

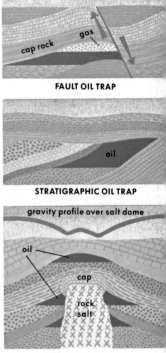

FAULT OIL TRAP

STRATIGRAPHIC OIL TRAP

gravity profile over salt dome

oil

cap

rock salt

SALT DOME OIL TRAP

99

ORE DEPOSITS are natural concentrations of metallic minerals in sufficient quantity to make their exploitation commercially worthwhile. Most industrial metals, other than iron, aluminum, and magnesium, are present as only a fraction of 1 percent in the average igneous rocks of the earth's crust. Platinum, for example, has an abundance of .000,000,5 percent by weight; silver and mercury, .000,01 percent. An ore with 1 percent antimony contains 10,000 times as much as the average igneous rock. There are several geological processes by which ore-bearing minerals are concentrated in the crust.

MAGMATIC ORES are concentrated from molten rock where crystals settle during cooling. Some of the world's greatest ore bodies were formed in this way.

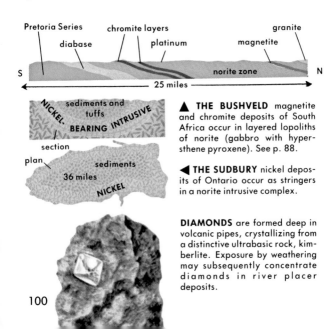

Pretoria Series — diabase — chromite layers — platinum — granite — magnetite — norite zone
S — N
— 25 miles —

sediments and tuffs — NICKEL-BEARING INTRUSIVE
section
plan — sediments — 36 miles — NICKEL

▲ **THE BUSHVELD** magnetite and chromite deposits of South Africa occur in layered lopoliths of norite (gabbro with hypersthene pyroxene). See p. 88.

◀ **THE SUDBURY** nickel deposits of Ontario occur as stringers in a norite intrusive complex.

DIAMONDS are formed deep in volcanic pipes, crystallizing from a distinctive ultrabasic rock, kimberlite. Exposure by weathering may subsequently concentrate diamonds in river placer deposits.

METAMORPHIC ORES are formed around some intrusions by contact metamorphism of the country rocks. Asbestos, once used for insulation and fireproofing, is a common nonmetallic metamorphic mineral.

HYDROTHERMAL ORES include many of the largest deposits of lead, zinc, copper, and silver. Deposited from hot, aqueous solutions, their distribution is usually controlled by joints, faults, bedding, and lithology of the country rocks. The mineralizing solutions arise from magmatic sources, although the parent igneous rock may not always be exposed. The copper deposits of Butte, Montana, and Utah, silver of Comstock Lode, Nevada, and gold of Cripple Creek, Colorado, are all hydrothermal deposits. A large portion of the world's gold is mined from deposits that originated from hydrothermal solutions.

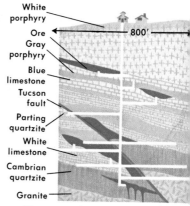

White porphyry
Ore
Gray porphyry
Blue limestone
Tucson fault
Parting quartzite
White limestone
Cambrian quartzite
Granite
800'

Diagrammatic cross section of structural ore control in lead and zinc mine, Leadville, Colorado (After Argall)

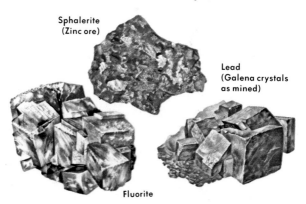

Sphalerite (Zinc ore)

Lead (Galena crystals as mined)

Fluorite

Drilling in taconite iron ore rocks requires special equipment.

METALLIC ORES, such as the iron around Birmingham, Alabama, and Northamptonshire, England, are sedimentary rocks rich in original hematite. Carnotite, a sedimentary uranium ore, is found in sandstones of the Colorado Plateau.

SEDIMENTARY ORES form in a variety of different ways. Some are formed as direct sedimentary deposits or by evaporation; others have an igneous origin and are concentrated by sedimentary processes.

EVAPORITES are mineral salts (halite, gypsum, potash, etc.) formed by evaporation of restricted bodies of water. The great salt deposits of Germany, Utah, and New Mexico are examples. Evaporites may be impure, containing clay, sand, or carbonates.

PLACER DEPOSITS are formed by concentration of fragments in stream deposits. Such heavy minerals as gold, diamonds, magnetite, and tin oxide are often concentrated in this way. Examples are the tin deposits of Malaysia and the goldfields of California and Australia.

RESIDUAL MINERAL DEPOSITS are formed by weathering and leaching, with corresponding enrichment of low-grade ore deposits, many of sedimentary origin. The great iron deposits of the Great Lakes and the aluminum (bauxite) deposits of Arkansas have been concentrated in situ by this process.

Iron ore pit, Virginia, Minn.; rocks are Precambrian.

CONSTRUCTIONAL AND INDUSTRIAL MINERALS,

although more abundant than ore minerals, are used in much greater quantities. They are generally quarried rather than mined.

BUILDING MATERIALS are extracted from the earth, whether brick, stone, steel girders, or glass. Building stones must be durable, easily quarried, and easily worked. The particular stones selected often depend upon the local weathering conditions (determined by amounts of industrial gases) and local availability of stones.

Other industrial minerals, such as sulphur, salt, potash, gypsum, and asbestos, are not so widely distributed, but are used in many industrial processes. They occur in very different geologic settings.

CLAYS are used for brick making, in chemical industries, in ceramics, and in the manufacture of many other products. Each use demands slightly different qualities. Many varieties of clay are known and mined.

SAND AND GRAVEL are widely used in concrete construction. Most supplies come from glacial and river deposits. Reserves are plentiful, but distribution is patchy.

STONE AGGREGATE for highway, airfield, and dam construction is also used in great quantities. It is resistant and cheaply quarried. In some tropical areas, it has to be imported.

LIMESTONE is used in the manufacture of cement, as a metallurgical flux, as an aggregate, and in agriculture. Limestones are widely distributed, and over 500 million tons are quarried annually in North America alone.

▼

THE CHANGING EARTH

If the processes of erosion and deposition were counteracted only by igneous activity, we should expect the continents to be featureless plains, broken only by active volcanoes or plateaus of lava. But there are considerable movements of the earth's crust.

Ancient caves, high above present beach, indicate former position of sea level, N. Ireland.

EARTHQUAKES provide dramatic evidence of crustal movements. Small relative movements of the crust, both vertical and horizontal, can very often be measured after earthquakes have taken place (p. 126).

RAISED BEACHES and buried forests often involve regional warping rather than simple uplift. Raised beaches show a rise in the California coastline over the past years, but on the other hand, parts of Denmark are sinking (see p. 62).

OTHER COASTAL FEATURES such as wavecut platforms indicate relative uplift of the land; drowned valleys indicate relative sinking (p. 69). Raised coral reefs often show uplift, tilting, and even reversal of movement.

FOSSILS of marine animals in land areas indicate relative uplift of the land. The presence of coal seams with land plants thousands of feet below present ground level shows that sinking has also taken place. Lake sediments are often interbedded with sediments that were deposited in the sea.

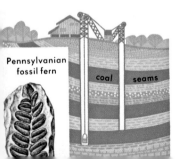

Pennsylvanian fossil fern

coal seams

Crustal movement is a common feature of the earth. It is found throughout the world, in rocks ranging in age from the oldest to the youngest. It involves various kinds of movement: gentle, slow uplift or sinking, regional warping, rapid earthquake movements, and regional stresses that are strong enough to buckle and break great masses of rock (pp. 107-125).

INCISED MEANDERS of rivers are thought to result from relative uplift of the land surface (p. 44). The Grand Canyon, which measures more than a mile deep, is an example.

RECENT MOVEMENT of the crust is indicated by famous temple ruins near Naples. The pillars, erected on land, were bored by marine molluscs, indicating submergence by the sea and the subsequent uplift of the land.

Roman temple, built on dry land, is now flooded.

Devonian sandstone

UNCONFORMITY

Silurian and Ordovician beds

UNCONFORMITIES (p. 114) represent breaks in the deposition of sediments, sometimes indicating periods of significant crustal disturbance.

Typical unconformity showing effects of uplift in pre-Devonian

Folding in limestone, Victoria, Australia▼

FOLDS in ancient strata range from small warps, only one or two feet in height, to great domes, miles across. In areas where rocks have responded in a brittle manner, breaks (faults) have developed.

FAULTS involve fracture and relative movement of rock units (p. 111). In nearly all cases, the rocks involved were originally in a horizontal position.

ROCK DEFORMATION

Although sedimentary rocks are generally deposited in almost horizontal beds, we generally find them distorted and tilted if we follow them over any considerable distance. Such structures are the result of large-scale crustal deformation, which produces corresponding changes in volume, shape, and sometimes chemical composition of the rocks themselves. The intensity of the changes is proportional to the intensity of deformation and the depth of burial (see Metamorphism, p. 94). Under some stress conditions, rocks behave as though they were elastic, but as stress increases, they undergo permanent (plastic) deformation and may ultimately fracture. The most intense zones of deformation are associated with mountain chains.

DIP of a bed is a measure of its slope or tilt in relation to the horizontal. The *direction of dip* is the direction of maximum slope, or the direction a ball would run over the bed if its surface were perfectly flat. The *angle of dip* is the acute angle this direction makes with a horizontal plane. The *strike* of a rock bed is the direction of the intersection of its dip direction with a horizontal plane. It is expressed as a com-

pass bearing and lies at right angles to the direction of dip. Dip-strike symbols are used on most geologic maps.

Clinometers are elaborate instruments used by geologists to measure dip. A simple instrument, however, may be made from a plastic protractor fixed to a flat base with a weighted thread to measure the maximum dip. The strike of the dip is then measured with a compass.

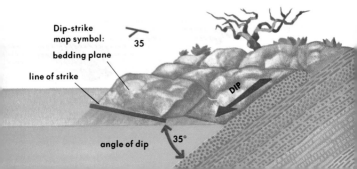

Dip-strike map symbol: 35

bedding plane

line of strike

DIP

angle of dip 35°

Sheep Mountain, Wyoming, is a fine example of pitching anticline.

FOLDS are wrinkles or flexures in stratified rocks. They range from microscopic sizes in metamorphic rocks to great structures hundreds of miles across. They sometimes occur in isolation, but more often they are packed together, especially in mountain ranges. Upfolds are called *anticlines*, and downfolds are called *synclines*. Folds with one limb more or less horizontal are called *monoclines*. All folds tend to die out as they are traced along their lengths.

FORMATION OF FOLDS shown in stages. In Stage 1, the rocks are deposited. In Stage 2, they are folded. In Stage 3, they are uplifted and their tops eroded, with only their dipping limbs remaining. The oldest beds (1) are always in the core of an anticline, but on the flanks of a syncline.

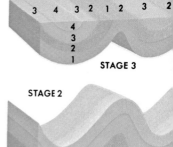

STAGE 3

STAGE 2

STAGE 1

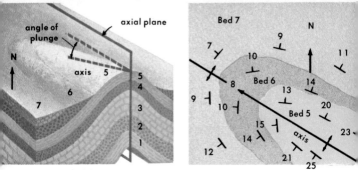

Relief diagram and geologic map of a plunging anticline

THE STRUCTURE OF FOLDS (above). The axial plane is drawn so that it bisects the angle of the fold. The axis is its trace on a bedding plane. If the axis is not horizontal, the fold is said to pitch or plunge.

ANTICLINES can be spotted on geologic maps by the dip arrows pointing away from the axis. A pitching fold gives a

SYNCLINES occur where the beds dip toward the axis. They can pitch and be either symmetrical or asymmetrical. A structural basin is a basin-shaped syncline where dips converge on a central point or area.

"closing" outcrop pattern. A dome is an anticline that has dips pointing in all directions from a central point or area.

KINDS OF FOLDS

SYMMETRICAL FOLDS (a) have limbs dipping in opposite directions at the same inclinations. The axial plane (ap) is vertical.

ASYMMETRICAL FOLDS (b) are those having an inclined axial plane and, like symmetrical folds, have limbs dipping in opposite directions, but at different inclinations.

OVERTURNED FOLDS (c) have an inclined axial plane and limbs dipping in the same direction. One limb is inverted.

ISOCLINAL FOLDS (d) have equal dips of two limbs; axial plane dips in same direction.

RECUMBENT FOLDS (e) have horizontal axial planes.

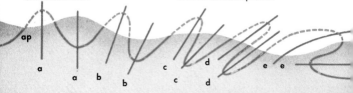

Folded rocks form cliffs 1,500 feet above San Juan River, Utah.

FOLD PATTERNS are rarely as simple as idealized ones shown on these pages. They often pass into faults. Some beds yield to strain more readily than others. Such incompetent rocks as shale and rock salt, for example, often yield by flowing and slipping in folds. Flowage of salt may produce structural domes, leading to petroleum reservoirs. Such incompetent rock movements are sometimes so strong that, as in some Middle East oilfields, major folds at depth are not reflected at the surface. Folds that do reach the surface are best exposed in arid and semiarid areas and in rock faces and cliffs, but regional mapping of other vegetation-covered areas, such as the Appalachians, often shows folding over great areas. Anticlinal fold structures sometimes provide petroleum reservoirs.

Section across southern Appalachian Mountains

| Cumberland Plateau | Ridge and valley belt | Blue Ridge | Piedmont Plateau | Carolina slate belt | Atlantic coastal plain |

NW   Metamorphic and plutonic belt   Triassic   SE

ROCK FRACTURES

JOINTS and fractures are another way that rocks yield to stress. Joints are fractures or cracks in which the rocks on either side of the fracture have not undergone relative movement. Common in sedimentary rocks, they are usually caused by release of burial pressure or by diastrophism. They play an important part in rock weathering as zones of weakness and water movement.

►**JOINTS IN SEDIMENTARY ROCKS** occur in parallel sets at right angles to the bedding. Tensional, compressional, and torsional stresses all produce distinctive joints.

JOINTS IN IGNEOUS ROCKS may result from shrinkage during cooling. In fine-grained rocks, there is a characteristic polygonal arrangement. Granite masses may show sheet jointing.

Giant's Causeway, Northern Ireland, shows hexagonal columns formed by cooling of basalt lavas.

FAULTS are fractures where once-continuous rocks have suffered relative displacement. The amount of movement may vary from less than an inch to many thousands of feet vertically and to more than 100 miles horizontally. Some, such as the San Andreas Fault, are major earth features. Different types of faults are produced by different compressional and tensional stresses, and they also depend upon the rock type and geological setting.

Faults are the cause of earthquakes, which suggests that repeated small movements rather than one "catastrophic" break characterize many faults. Distinctive, large-scale fracture zones (transform faults) displace the mid-oceanic ridges in several areas (pp. 136 and 140).

KINDS OF FAULTS

NORMAL, GRAVITY, OR TENSIONAL FAULTS are not necessarily the most common fault type in a given area. They are faults in which relative downward movement has taken place down the upper face or hanging wall of the fault plane. (We cannot generally prove whether both beds have moved, or only one.) The throw of the fault is the vertical displacement of the bed (ac); the heave is the horizontal displacement (bc). The angle abc is the dip of the fault plane, and the complement of this is the hade. The dip is usually steep. Sometimes there may be more than one episode of movement along the same fault plane or zone.

Block diagram of normal fault before weathering

Block diagram of normal fault after weathering

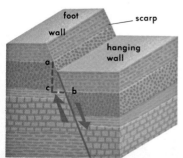

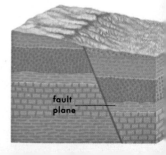

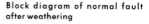

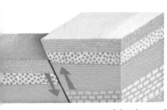

Block diagram of high-angle reverse fault

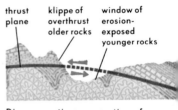

Diagrammatic cross section of eroded low-angle thrust fault

REVERSE OR THRUST FAULTS have relative upward movement of the hanging wall of the fault plane. They occur in areas of compression and folding such as mountain belts. Lateral displacement may be many miles. They often have a low dip and result in repetition and apparent reversal of stratigraphic order in a vertical sequence. Chief Mountain in Montana is an eroded remnant of a large thrust fault.

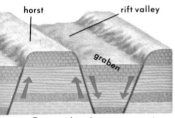

Topographic depression marking San Andreas Fault is occupied by a lagoon, Bolinas Bay.

A GRABEN is a block that has been dropped down between two normal faults. An uplifted fault block is a *horst*. *Rift valleys* are graben structures hundreds of miles in length. The most spectacular is that along the Red Sea, but they are also found in East Africa, the Rhine, and California, and they also occur beneath the oceans along the crests of the mid-oceanic ridges.

TEAR FAULTS, OR STRIKE-SLIP FAULTS, AND RELATED TRANSFORM FAULTS are those where shearing stress has produced horizontal movement. The San Andreas Fault in California is 600 miles long and has a displacement of over 350 miles. The 1906 and 1989 San Francisco earthquakes were caused by the movement of the San Andreas Fault, which is still active.

FAULTS IN THE FIELD are generally more complex than those shown in these diagrams. Rotational movements often complicate the simple vertical and horizontal movements, and the throw and hade of a fault may change along its length. The fault plane is often poorly defined, and is represented by a fault zone made up of broken and distorted rocks. Faults often occur in groups (fault zones) made up of many individual faults. Except in desert areas, cliff faces, and quarries, faults are rarely seen at the surface, but their presence is indicated by one or more of the following features:

FAULT BRECCIA occurs where the rocks of the fault zone are shattered into angular, irregularly sized fragments. Some may be reduced to a gritty clay.

TOPOGRAPHIC EFFECTS occur when faults bring together rocks of differing hardness. Denudation may indicate the faulting by showing sharp, "unnatural" topographic contact. The Teton range in Wyoming is an example, where resistant Precambrian igneous rocks are faulted against softer Tertiary sediments (p. 118). Other topographic effects—bays or valleys, for example—may result from the weakness of a fault zone, which itself may undergo strong differential weathering.

SPRINGS may be produced by faults when pervious and impervious strata are brought into contact with one another (p. 51). Lines of springs often indicate the existence of a fault.

SLICKENSIDES, polished striations or flutings, are often found in the fault plane or zone. They may indicate the direction of relative movement.

DISPLACEMENT OF OUTCROP is another indication of a fault. As well as displacing rocks vertically, faults in dipping beds will displace their outcrop patterns. In geologic mapping, faults are often inferred from outcrops that "won't match."

Small-scale faulting in limestone

UNCONFORMITIES are ancient erosion surfaces in which an older group of rocks has been uplifted, eroded, and subsequently buried by a group of younger rocks. Unconformities can be used to date periods of crustal movement in the reconstruction of earth history. The time span represented by the erosion surface (absence of deposits) may vary from very long to very short periods. A "break" in fossil sequence, often a characteristic feature of unconformities, can be measured by comparing the fossils with those of uninterrupted rock sequences. Unconformities vary in form, and in the erosional intervals they represent.

DISCONFORMITIES occur where the beds above and below the erosion surface are parallel. If the older group of rocks has been folded and eroded before the deposition of the younger group, their beds will not be parallel, and they represent an *angular unconformity*.

The eroded upper surface of the underlying group of rocks may be flat or it may have considerable relief. Old soils are sometimes preserved. The lowest overlying beds are often *conglomerates* made up of eroded fragments of the underlying group of rocks.

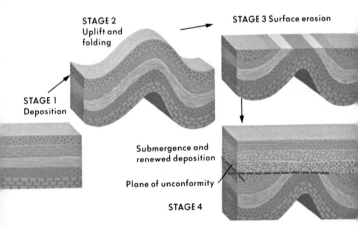

STAGE 2
Uplift and folding

STAGE 3 Surface erosion

STAGE 1
Deposition

Submergence and renewed deposition

Plane of unconformity

STAGE 4

The Matterhorn, Switzerland, shows sharp outline formed by glacial erosion typical of many mountains.

MOUNTAIN BUILDING

Mountains are among the most conspicuous features of the earth's surface. Although they are formed in various ways and are of various ages, most are concentrated in great folded belts that run along the continental margins. Not all are confined to the land. The Mid-Atlantic Ridge stretches thousands of miles, rising almost 10,000 feet above the ocean floor. Similar ridges are found in other oceans (p. 136).

Any isolated, upstanding mass may be called a mountain. There is no minimum height or particular shape involved. Some of the ways in which mountains may be produced are shown on the following pages.

VOLCANIC MOUNTAINS include some of the world's most beautiful and famous mountains. Mt. Fujiyama in Japan, Mt. Vesuvius in Italy, Mt. Hood in Oregon, and Mt. St. Helens and Mt. Rainier in Washington are all examples of characteristically steep and symmetrical volcanic mountains. Other volcanic mountains, such as Mauna Loa in Hawaii, are formed by shield volcanoes (p. 84). They tend to be rounded and flattened, but may still be high and large. Mauna Loa rises almost 14,000 feet above sea level and about 32,000 feet above the seafloor. With its diameter of 60 miles, it is the world's largest mountain in terms of volume.

Volcanic mountains may form very rapidly. Mt. Paracutin in Mexico began to grow in February, 1943. Within a week, its cone was 500 feet high, and within two years it had reached 1,500 feet.

Some volcanic mountains occur as isolated structures, but others form parts of extensive volcanic chains. Volcanic islands may form great island arcs, of which the 1,000-mile Aleutian chain is an example.

Mt. Fujiyama, Japan's sacred mountain, is a volcano.

Blue Mountains, New South Wales, Australia, are erosional in origin, consisting chiefly of horizontal Permian and Triassic strata.

EROSIONAL MOUNTAINS are found in regions of crustal uplift. These are wide areas where crustal uplift (epeirogeny) has produced elevated plateaus and thereby provided new erosional energy to rivers. The steep gorges with precipitous edges of Grand Canyon-type topography, for example, are generally thought of as a dissected plateau rather than a mountain. Some well-known mountains, such as the Blue Mountains of New South Wales, Australia, are formed in this way. If erosion continues long enough, hugh isolated, resistant remnants are left. An example is Mt. Monadnock in New Hampshire, which rises 1,800 feet above a "peneplane" and gives its name to such isolated structures (monadnocks). These structures are thus outliers of younger rock, resting on an exposed basement of older rocks.

117

DOME MOUNTAINS are the simplest kind of structural mountains. They tend to be relatively small, isolated, structural domes, uplifted without intense faulting. They may be associated with such igneous intrusions as laccoliths. The Black Hills of South Dakota are an example.

STRUCTURAL MOUNTAIN RANGES are by far the most numerous and extensive mountain ranges and differ from volcanic and erosional mountains in the structural deformation and uplift they have undergone. Some of the older mountain ranges, such as the Appalachians, are less impressive than the younger, such as the Alps, but they share certain common characteristics, six of which are discussed below. It is these, more than just height, that are a clue to an understanding of the nature of structural mountains.

FAULT BLOCK MOUNTAINS may be formed in any type of rock. They are often, though not always, structurally undeformed where differential vertical displacement or tilting by steep normal faults gives relatively uplifted and downsunk blocks. Rift valleys are often associated with the mountain scarps, which tend to run parallel with the faulting. Such mountains may be very large. The Sierra Nevada represents the uptilted edge of a block of granite 400 miles long and 100 miles wide, with an eastern scarp 13,000 feet above sea level. There are several such ranges from Oregon to Arizona, which produces spectacular scenery.

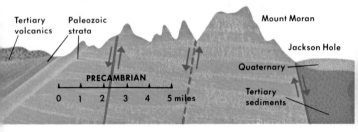

Tertiary volcanics

Paleozoic strata

Mount Moran

Jackson Hole

PRECAMBRIAN

Quaternary

0   1   2   3   4   5 miles

Tertiary sediments

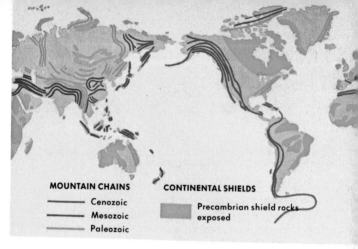

FOLDED MOUNTAIN RANGES all have similar character-istics. Most of the world's great mountain chains, such as the "young" Himalayas, Alps, Rockies, and Andes, and the older Urals and Appalachians, belong to none of the groups discussed on pp. 116-117. In spite of many individ-ual differences, all are fundamentally folded ranges and share these six important characteristics:

1. Linear distribution of mountain ranges indicates they are not haphazardly scattered across the earth's surface; they are found in long, narrow belts. The Appalachians, for example, stretch 1,500 miles from Newfoundland to Alabama and are up to 350 miles wide. The Rockies are 3,000 miles long and continue southward for another 5,000 miles in the Andes of South America and then into Antarctica. Their location near the margins, or former margins, of continents is a major clue to their origin.

119

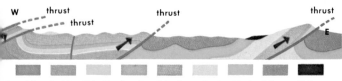

W   thrust   thrust   thrust

thrust   E

Precambrian   Cambrian   Ordovician   Devonian   Carboniferous   Permian   Triassic   Cretaceous   Tertiary

Generalized section across the Rockies from eastern Idaho to western Wyoming (see p. 152 for meaning of geologic ages)

2. Thick, sedimentary rocks that are found in all the world's major mountain chains are generally 30,000 to 40,000 feet (6-8 miles) in thickness. Fossils and structures show most are marine rocks, formed in elongated crustal downwarps. They include chiefly clastic sediments and volcanic rocks. Strata of equivalent age on the bordering continents are often only a tenth of this thickness and generally have a higher proportion of carbonates and coarse clastic rocks.

3. Distinctive igneous and metamorphic rocks are found in mountain chains. Clastic sediments are often interbedded with great thicknesses of volcanic lavas, and eroded cores of mountain chains include vast granitic batholiths, as well as migmatites. The batholiths and migmatites were probably formed from the effect of depth and heat on deeply buried sediments.

In younger mountain ranges—those formed during the last 120 million years (see pp. 119 and 152)—active vulcanism continues. The volcanoes of the Cascades, Andes, and Antarctica are examples. In older mountain ranges, although the elevation seen in more recent mountains is reduced by erosion, the distinctive rock types and deformation patterns can still be recognized.

120

4. Intense folding and faulting in major mountain chains often involve thrusting on the neighboring continental margin, with displacements of many miles and consequent shortening of the crust. Active earthquakes and uplift mark "younger" mountain ranges.

5. Repeated uplift and erosion have taken place in folded mountains. Older ranges may be eroded down almost to a flat surface (a peneplain) and then be uplifted again. The Appalachians, formed in the Late Paleozoic, were eroded to a peneplain in the Triassic and again in the Cretaceous. Their present 6,000-feet-high relief results from Tertiary uplift and recent erosion. Continental shield areas, now rather flat, represent ancient eroded mountain chains.

6. Mountains have roots. They are not simply lumps of rock resting on a uniform surface. They have roots of light material extending far down below their normal depths. This is suggested by the gravity anomaly of mountains, by the paths of earthquake waves through them, and by high heat flow associated with mountains. Since mountains deflect a plumb line less than the mere attraction of their surface mass resting on "normal" basement, they must be underlain by a larger amount of light material.

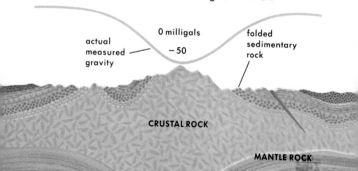

0 milligals

actual measured gravity

−50

folded sedimentary rock

CRUSTAL ROCK

MANTLE ROCK

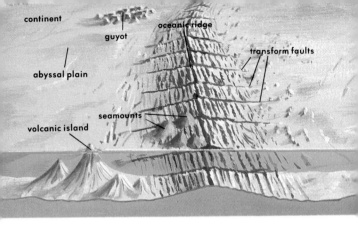

Main features of the ocean floor

THE ARCHITECTURE OF THE EARTH

SEEN FROM SPACE, earth is a blue planet, veiled in a swirling, changing tracery of clouds. It is blue because it is a watery planet, two-thirds of its surface covered by oceans; it is veiled because its enveloping atmosphere is in constant motion, interacting everywhere with land and water. The first part of this book has described this interaction, showing the changing pattern of rock formation, erosion, and deposition. But beneath these surface changes, there are also changes within the earth, which have created its major architectural features. The second part of this book describes these major earth features and the processes that produced them.

The continents consist of large shields of ancient Precambrian rocks, around which younger rocks have been formed.

Mountain ranges of several different ages are found in linear belts around the margins of the continents. They are

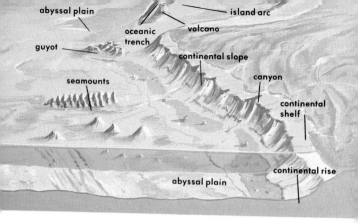

abssyal plain — island arc — oceanic trench — volcano — guyot — continental slope — seamounts — canyon — continental shelf — abyssal plain — continental rise

(After *The Story of the Earth*, H.M.S.O.)

intensely folded and faulted and contain a variety of sedimentary, igneous, and metamorphic rocks (p. 119).

Volcanoes and *earthquakes* are not randomly distributed, but are concentrated in narrow belts, especially in a "ring of fire" around the Pacific Ocean, in areas of recent mountain building, and along the mid-ocean ridges and rift valleys (p. 85).

Mid-ocean ridges form a global network of submarine mountain chains, 25,000 miles long and reaching 18,000 feet in height above the seafloor. Rift valleys, transform faults, shallow-focus earthquakes, and volcanoes mark the crest of the mid-ocean ridge (p. 136).

Island arcs and *submarine trenches*, some over 6 miles deep, found in the Pacific and elsewhere, are marked by vulcanism and intense earthquake activity (pp. 138-139).

The ocean floor is made up of volcanic and sedimentary rocks, all of relatively young age. In contrast to the ancient rocks of the continents (some 3.6 billion years old), the oldest oceanic rocks are "only" 175 million years old.

THE MECHANISM OF MOUNTAIN BUILDING (ORO-GENY) is one of the great discoveries of recent geological sciences. Although surface processes play a significant role in supplying the sediments that make up mountain ranges and in eroding and accentuating them after they are uplifted, the chief processes operate within the earth's interior. The following pages trace the way in which recent discoveries have led, step-by-step, to a new theory of mountain building—plate tectonics—that accounts for the features just described. To understand this theory, we need information on the structure of the earth and its physical properties. The following pages describe these features.

EARTHQUAKES

Earthquakes are rapid movements of the earth's crust caused by fault movements. Almost a million occur each year, but most are so weak they may be detected only by very sensitive recording instruments (seismographs). About 90 percent of all earthquakes seem to originate at a focus, the point of maximum intensity of the earthquake, in the outer 40 miles of the crust. A few deep-focus earthquakes originate at depths as great as 400 miles. The location and pattern of earthquakes are major clues to earth's structure.

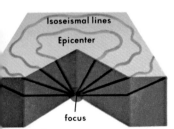

TYPICAL EARTHQUAKE EFFECTS are shown on block diagram. Earthquake waves radiate outwards. Those that travel directly to the surface are shown. Isoseismal lines join all places where the earthquake is recorded with the same intensity. These generally form approximate circles or ovals of decreasing intensity around the epicenter, which is directly above the focus.

Destruction caused by the California earthquake of October, 1989

EARTHQUAKE DESTRUCTION occurs when shock waves are absorbed by buildings. Those on thick soil or rock debris are most affected. Steel-frame buildings on solid rock are more likely to survive. Violent earth movements may be accompanied by earth flows and slumps, cracks, small faults (with horizontal or vertical movement up to about 30 feet), and temporary fountains. Fires often result from broken gas lines.

Tsunamis are giant waves in the oceans that result from earthquakes. Movement of a part of the ocean floor disturbs the water and leads to huge oscillatory waves, often over 100 miles long. Moving at speeds of up to 500 mph, they pile up along shores, often causing great destruction.

EARTHQUAKE BELTS encircle the earth. About 4/5 of all earthquakes and almost all deep-focus ones occur in a belt around the Pacific. A less active belt runs through the Himalayas and Alps. (Compare with p. 145.)

Circum-Pacific

Mediterranean and Trans-Asiatic

Mid-Atlantic, Mid-Indian, and East African

MAJOR EARTHQUAKE BELTS

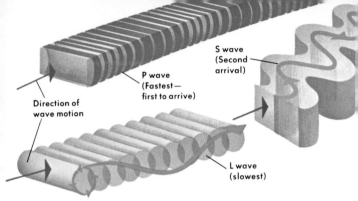

EARTHQUAKE RECORDS, or seismograms, show that various types of earthquake waves travel through the earth by different paths. There are two types of waves that travel *through* the earth: P (push-pull) waves that are compressional and travel by moving the particles backward and forward through any material, and S (shake) waves that pass only through solids. S-waves shake the particles perpendicularly to the direction of their movement, like waves made in a rope by holding one end still and shaking the other up and down. L (surface) waves travel around the

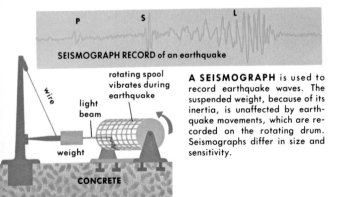

SEISMOGRAPH RECORD of an earthquake

A SEISMOGRAPH is used to record earthquake waves. The suspended weight, because of its inertia, is unaffected by earthquake movements, which are recorded on the rotating drum. Seismographs differ in size and sensitivity.

surface of the earth and can pass through any material. From many observations, it has been shown that when recording stations are more than 7,000 miles away from an earthquake epicenter, the otherwise very regular travel times of P-waves are reduced. They arrive later than calculated, and S-waves are eliminated completely. This occurs because changes of composition within the earth's interior influence the paths and speeds of the various waves.

By using a worldwide seismograph network and studying the paths and travel times of these waves, it is possible not only to locate an earthquake at its source but also to build up a model or picture of the nature of the earth's interior.

EARTHQUAKE WAVES passing through the earth are illustrated below, showing the effect of refraction on P-waves. S-waves are completely eliminated by the core. P-waves are indicated by red lines, S-waves by the blue lines.

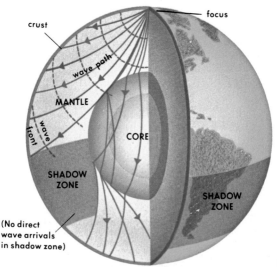

THE EARTH'S INTERIOR

The behavior of earthquake waves indicates that the earth's interior is not homogeneous but is made up of a series of concentric, layered shells of different composition.

Using data from seismographs, a hypothetical model of the interior shells can be constructed. We have no direct method of testing its accuracy, but there are several lines of indirect evidence that seem to support it (p. 129). The three main layers are called crust, mantle, and core. The crust is separated from the mantle by the Mohorovicic, or M-discontinuity, where earthquake waves speed up by 15 percent.

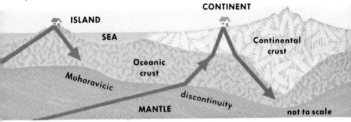

ALTERNATIVE CLASSIFICATION recognizes the rigid *lithosphere* (the outer 62 miles), consisting of the crust and the upper part of the mantle. This overlies the hot, soft *asthenosphere* (62-155 miles down), characterized by low seismic wave velocities and high seismic attenuation, which is thought to be capable of flowing. This overlies the stronger *mesosphere*, the lower mantle. It appears that the major plates of the earth rest on and move across the asthenosphere (see p. 144).

THE CONTINENTAL CRUST is thicker than the oceanic (20-38 miles, as opposed to 5 miles), older (up to 4 billion years, as opposed to a maximum of 175-200 million years), lighter (2.7 gms per cc, as opposed to 3.0 gm/cc), and more structurally complex. The continental crust has a granitic composition, with thick sedimentary and metamorphic rocks, while the oceanic crust is basaltic, with only a thin veneer of sediments. These differences are explained by the theory of plate tectonics.

THE MANTLE, bounded by the Mohorovicic discontinuity, is marked by 15-percent increases in P- and S-wave velocities, suggesting a change in composition. It seems to consist of dark, heavy rocks, rich in iron-magnesium silicates (olivine and pyroxene, pp. 30 and 89). A few surface outcrops of supposed mantle rocks fit this general composition. Slow convection movements in the upper mantle influence the structure of the crust, allowing the movements involved in plate tectonics.

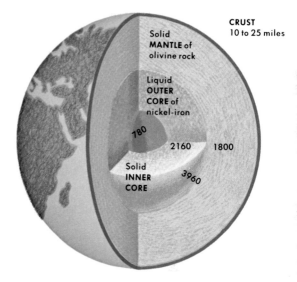

CRUST
10 to 25 miles

Solid **MANTLE** of olivine rock

Liquid **OUTER CORE** of nickel-iron

Solid **INNER CORE**

780

2160

1800

3960

THE CORE is recognized as a major discontinuity within the earth where the S-waves are eliminated and P-wave velocity is slowed down. This indicates that the outer core is probably a "liquid," but the fact that P-waves speed up again at a depth of 3,160 miles suggests that the inner core is solid. Circulation movements in the liquid outer core probably generate the earth's magnetic field (p. 135). It probably has an iron-iron sulphide composition because of its high density (15 for the inner core), and by analogy with meteorites, which probably formed at the same time. It may, however, be a similar composition to the mantle, its differences arising from variations created under very high pressure.

GEOPHYSICAL MEASUREMENTS provide additional information about the earth's interior and mountain-building processes. Artificial shock waves, used in seismic studies of subsurface structure, reveal the nature and boundaries of shallow layering within the earth's interior. Measurements of gravity and magnetic variations are useful in subsurface studies, both "deep," as in those beneath mountain roots, and "shallow," as in the location of petroleum traps or mineral deposits. Heat-flow studies are used to analyze deep structures. Radar mapping is also of importance. Continuous, rapid geophysical survey measurements are now being made from aircraft and ships.

Heat flow related to distance from Mid-Atlantic Ridge. Note high heat flow at axis.

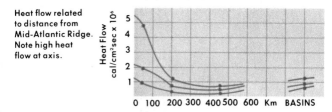

HEAT-FLOW STUDIES in mines and boreholes show that the temperature of the earth's crust increases by an average of 30°C per kilometer of depth, although there are wide local variations caused by the geological setting and local conductivity. Over the continents, most of the background heat seems to originate from radioactivity in the crust. Where the crust is thick, as in mountain ranges, values tend to be high.

Seismic shot explosion produces a plume of soil. Seismic studies also use "thumper" techniques, without explosives.

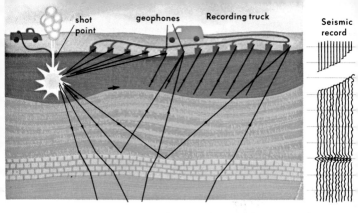

shot point · **geophones** · **Recording truck** · **Seismic record**

GEOPHYSICAL EXPLORATION is widely used in the location of mineral deposits and petroleum traps, and in geologic mapping. The diagram shows how seismic studies provide a structural profile of layered rocks. Geophones placed at measured intervals record the trace and time of P-waves.

ELECTRIC LOGGING of the resistivity and self-potential of rock penetrated by boreholes is measured and plotted. Electric logging is an aid in subsurface correlation of oil wells. (Landes)

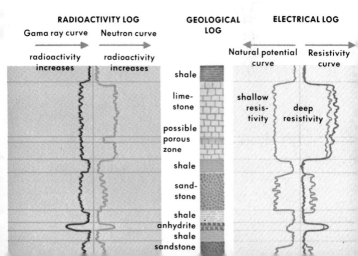

RADIOACTIVITY LOG

Gama ray curve → · Neutron curve →

radioactivity increases · radioactivity increases

GEOLOGICAL LOG

shale
lime-
stone
possible
porous
zone
shale
sand-
stone
shale
anhydrite
shale
sandstone

ELECTRICAL LOG

← Natural potential curve → · ← Resistivity curve →

shallow resistivity · deep resistivity

THE FORCE OF GRAVITY is present in every part of the universe, from the smallest particle to the largest star. The earth attracts everything around it with a pull towards its center. The sun holds its planets in their orbits by gravitational attraction. The force (F) involved is not constant but varies inversely with the square of the distance between the two bodies involved (D), and directly with their masses (M_1 and M_2). We can calculate it by using the equation developed by Newton, $F = \dfrac{GM_1M_2}{D^2}$, where G is the earth's gravity constant (6.67×10^{-11} cubic meters per kilogram second^2).

The force of gravity is not the same at every place on the earth. It is lower on high mountain tops, and shows a general decrease of about one-half of one percent from the poles toward the equator.

Gravity influences almost all geological processes. Weathering and erosion, patterns of sediment distribution, the form of mountains, and even the cooling of igneous rock, all reflect the influence of gravity.

Principle of gravity is shown by spring balance; relative extension of spring is proportional to density of underlying rock.

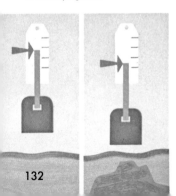

A GRAVIMETER is used to make rapid and accurate determinations of relative differences in the earth's gravity field. It is a very sensitive spring balance, an increase in gravity being recorded by the stretching of the spring. Gravimeters employ optical and electrical methods to "magnify" the minute increase in spring length that has to be measured.

GRAVITY PROFILE may also show low gravity (negative anomaly) due to presence of lightweight salt in a salt plug, invisible from the surface (see p. 99).

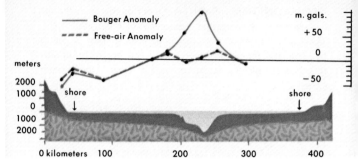

THE RED SEA AND GULF OF ADEN, unlike other rift valleys, have positive gravity anomalies. They are underlain by basic rocks and are bounded by parallel faults. This suggests they were formed by the crustal separation of Arabia and Africa, the "gap" between them being filled by rising basic material from the mantle.

ISOSTASY is the state of equilibrium that exists in the earth's crust. Because mountains have roots (p. 121) and also stand above the average level of the generally similar rocks of surrounding areas, a balance must exist between them and the denser material on which they

MOUNTAIN CHAINS often have a major negative gravity anomaly caused by a root of lightweight granitic material. Negative anomalies across rift valleys are partly explicable by thick sediments within the valleys themselves, and may result from the concentration of lightweight alkaline magmas and volcanoes in rift areas.

"float." As erosion reduces the mass of the mountain, uplift takes place below it as plastic material flows under it—just as unloading cargo from a ship causes it to rise in the water. The postglacial uplift of Scandinavia is an example of isostatic adjustment (p. 62).

Airy's isostatic hypothesis (1) suggested equilibrium results if crustal blocks of similar density have different heights; Pratt's hypothesis (2): blocks of different density have a uniform level of compensation. Modern hypotheses suggest variation in density in and between columns.

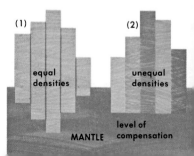

GEOMAGNETISM reflects the earth's behavior as though it were a giant bar magnet surrounded by a magnetic field. The force causes a compass to rotate so that it points towards the magnetic north pole. The earth's magnetic field is probably caused by convection currents in the outer core.

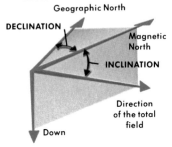

MAGNETIC DECLINATION is the angle between geographic ("true") north and magnetic north. The vertical angle between the horizontal and a freely dipping magnetic needle is the *inclination*. The declination shows daily changes and also slow, measurable changes over long periods of time. The magnetic poles change position relative to the geographic poles at a present rate of about four miles every year, although the deviation is never large.

MAGNETIC STORMS are sudden fluctuations in the earth's magnetic field caused by charged particles from the sun (the solar wind). Magnetic storms often precede aurora displays.

MAGNETOMETERS measure the local intensity of the earth's magnetic field. Variations (anomalies) are caused by rocks of differing magnetic properties. Magnetic traverses can be made on the ground or by airborne or seaborne magnetometers. Regional surveys are used in mineral exploration and in study of the ocean floor.

PALEOMAGNETISM is the remnant magnetism found in rocks—especially lavas and some sedimentary rocks—reflecting the ancient magnetic fields at the time of their formation. Magnetic particles in the rocks orient themselves like compass needles, reflecting the field in which they formed.

Airborne magnetometer

THE MAGNETIC FIELD of the earth has been measured and mapped. The lines of force, marked by arrows, show its direction at various places. It also varies in intensity (being twice as great at the poles as at the equator), and the inclination ranges from 0° at the magnetic equator to 90° at the magnetic poles. Local anomalies are often produced by distinctive magnetic properties of some rock bodies.

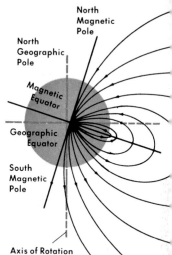

REVERSALS in earth's magnetic field have occurred every few hundred thousand years during the last 70 million years. A time scale of reversals has been reconstructed from sequences of lavas and deep-sea sediments.

MAGNETIC SURVEYS OF OCEAN FLOOR show a distinctive parallel pattern of magnetic reversal "anomalies" across the mid-ocean ridges, and provide a clue to the history of the ocean (p. 140).

POSITIONS OF ANCIENT CONTINENTS are reconstructed by measurements of remnant magnetism in rocks of successive ages. For Europe and North America, the traces of ancient polar positions are similar, but are displaced parallel to one another, suggesting either that the poles have wandered or that the continents have moved apart. Other independent evidence supports the interpretation of major continental movements.

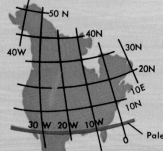

TRIASSIC EQUATOR in North America based on paleomagnetic data (after Irving). Zero paleomeridian is arbitrarily taken through New York. Paleomagnetic data show North America was once joined to Europe, Africa, and South America within a single continent, Pangea, which began to split apart about 200 million years ago.

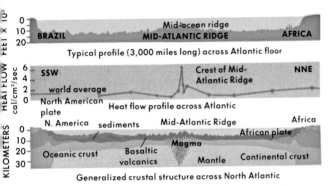

Typical profile (3,000 miles long) across Atlantic floor

Heat flow profile across Atlantic

Generalized crustal structure across North Atlantic

STRUCTURE OF THE OCEAN FLOOR

MID-OCEAN RIDGES form a global network of mountain chains, 25,000 miles long, up to 1,500 miles wide, and up to 18,000 feet high (p. 66), formed of young volcanic rocks. They are repeatedly offset by transform faults and have crests marked by rift valleys, up to 12,000 feet deep and 30 miles wide. In some places (e.g., Iceland and East Africa) ridges emerge above sea level.

MID-OCEAN RIDGES are marked by abnormally high heat flow, shallow earthquakes, volcanic activity, rift valleys, and transform faults. All these imply that the ridges are places of strong crustal tension. Geophysical studies, dredging, and drilling show the ridges have a thin veneer of deep-sea sediments, but consist chiefly of basaltic pillow lavas, overlying vertical feeders, and deep gabbroic crystalline rocks. All these features imply that the ridges are sites for the submarine extrusion of new crustal material (see pp. 140-141).

Mid-Ocean Ridges

Trans-
form
Faults

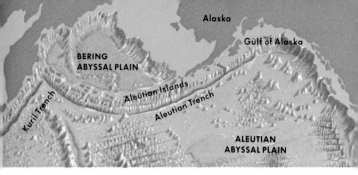

ABYSSAL PLAINS cover over 60 percent of the ocean floor. They are flat basins, hundreds of miles across, generally broken only by ocean ridges, canyon systems, trenches, volcanic islands, or seamounts. They lie at depths below 15,000 feet. Geophysical studies show that they have a thin cover of young sediments.

ROCKS OF THE OCEAN FLOOR are younger than the continents, none being older than 175 million years; some continental rocks are over 20 times that age. Ocean sediments thicken and include successively older layers in parallel bands away from the mid-ocean ridge. Underlying submarine lavas show the same feature (p. 140). These observations are important in the development of a model to explain the history of the earth.

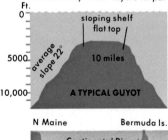

VOLCANIC SEAMOUNTS rise up to 12,000 feet above the abyssal plains. Guyots are similar, but have flat tops, presumably having been eroded by wave action and later submerged. Most lie at 3,000-5,000 feet below present sea level.

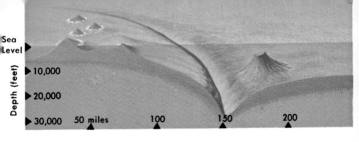

Sea Level ▶

Depth (feet)
▶ 10,000
▶ 20,000
▶ 30,000

50 miles ▲   100 ▲   150 ▲   200 ▲

VOLCANIC ISLAND ARCS, common around the Pacific Basin, are major earth features, generally bordered on their convex oceanic side by narrow trenches, up to 36,000 feet deep. They are sites of vulcanism, earthquakes, negative gravity anomalies, and crustal instability. Their andesitic lavas, intermediate in composition between those of the continents and oceans, suggest a mixture of the two. Nearly all deep-focus earthquakes occur below these arcs. Geophysical studies (p. 139) indicate that island arcs reflect the collision of two plates of the earth's crust. The trace of earthquake foci marks the buckling under (subduction) of one plate. The islands seem to be formed from melted material from the subducted plate rising through the overriding plate, forming volcanoes and intrusions. Japan and the Aleutian and Marianas islands are examples. Destruction of "old" ocean crust by subduction (p. 139) balances new crust formed at the ridges.

TRENCHES OF THE PACIFIC BASIN

138

1. Aleutian
2. Kurile
3. Japan
4. Nansei Shoto
5. Mariana
6. Palau
7. Philippine
8. Weber
9. Java
10. New Britain
11. New Hebrides
12. Tonga-Kermadec
13. Peru-Chile
14. Acapulco-Guatemala
15. Cedros

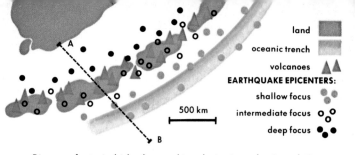

Diagram of a typical island arc and trench structure, showing relative position of volcanoes; shallow-, intermediate-, and deep-focus earthquakes; and oceanic trench. AB represents line of sections below; length about 1,200 kilometers.

TOPOGRAPHIC PROFILE shows very steep sides of typical island-arc trench.

GRAVITY PROFILE shows island-arc trench areas as belts of strong negative anomalies.

HIGH HEAT FLOW in volcanic areas but low across trenches reflects "cool" oceanic plate being melted and uprising.

EARTHQUAKE EPICENTER PLOT, dipping toward the continents at 30°-60°, reflects subduction of oceanic plate.

MODEL OF ISLAND ARC shows how subduction of plate creates trench, earthquakes, volcanoes, and heat and gravity features.

(After *The Story of the Earth*, H.M.S.O.)

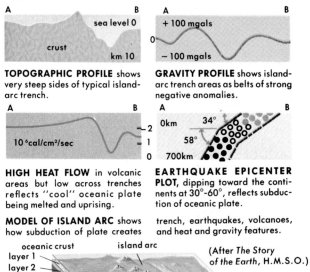

139

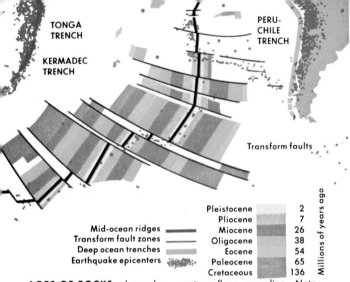

TONGA TRENCH

KERMADEC TRENCH

PERU-CHILE TRENCH

Transform faults

Mid-ocean ridges	
Transform fault zones	
Deep ocean trenches	
Earthquake epicenters	

		Millions of years ago
Pleistocene		2
Pliocene		7
Miocene		26
Oligocene		38
Eocene		54
Paleocene		65
Cretaceous		136

AGES OF ROCKS, shown by magnetic anomalies, existing as mirror images across the Pacific mid-oceanic ridge, strongly suggest seafloor spreading. Note concentration of earthquakes; deeper focus ones are confined to trench areas shown.

SEAFLOOR SPREADING is a mechanism that accounts for the major features of the continents and oceans. New ocean floor is created along the length of the mid-ocean ridges, from which it then moves away in both directions at rates of between 1/2 inch and 3 inches a year. This seems to be part of a larger movement in which the outer shell of the earth (the lithosphere) glides slowly across the molten upper layers of the mantle (the asthenosphere, p. 128), which acts as a conveyor belt for the rigid plates that include both oceans and continents. Both the basaltic rocks of the ocean crust and the overlying sediments become successively older away from the ridges, but all are less than 175 million years old.

PRESENT OCEAN BASINS are young features of the earth, earlier ocean basins having been destroyed by plate tectonic movements in subduction and collision.

TRANSFORM FAULTS (p. 140) offset the mid-ocean ridges. They are probably caused by different spreading rates along the ridge. Earthquakes occur only along parts of faults between the two ridge segments, because spreading movement across other parts of fault is in same direction. The San Andreas Fault is a transform fault that outcrops on land, with the Pacific Plate sliding at a rate of 4-6 cm. a year past the American Plate. Sudden movements along this fault produce destructive earthquakes.

RATE OF SPREADING from mid-ocean ridge is measured from symmetrical pattern of magnetic reversals observed in rocks of equal age. These reversals are dated by studies of similar reversals in lava sequences. The distance of bands of equal age from the ridges shows differences in rate of spreading. The East Pacific is opening up at about 12 cm. (¾ in.) a year, the North Atlantic by only about 2 cm. a year.

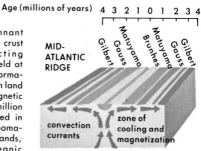

ANOMALIES in the remnant magnetism of the oceanic crust show reversals, reflecting changes in the magnetic field at the various times of crust formation. Studies of lava flows on land give a time scale of magnetic reversals for the past 4 million years, and this is confirmed in deep-sea cores. Similar anomalies exist in symmetrical bands, parallel to the mid-oceanic ridges. This suggests that new crust is being formed and spreading at rates up to 12 centimeters per year.

Model shows rise of new seafloor, "imprinted" with reversals of successive magnetic periods and reversals.

SUBMERGED VOLCANOES are scattered across the ocean floor, either singly or in clusters, with more than 10,000 in the Pacific. Many formed above sea level, but have since been submerged by plate movements (p. 144).

THE PRATT-WELKER chain of guyots runs from the Gulf of Alaska toward the Aleutian Trench (p. 137). Guyot GA-1 has a surface lying almost 6,000 feet below the general level of all the other guyots.

It has been carried downwards by the subsidence that produced the Aleutian Trench.

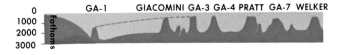

SOUTHEASTWARD MIGRATION of volcanic activity in several island chains is explained by a stationary plume or hot spot, rising from the mantle through a northwest moving plate.

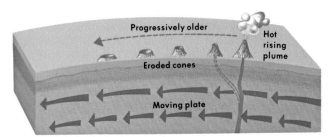

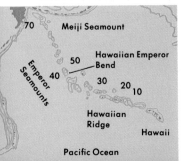

HAWAIIAN ISLAND-EMPEROR SEAMOUNT chain has active vulcanism limited to the island of Hawaii at the extreme southeast tip. Extinct volcanic islands are progressively older towards the northwest. Hawaiian Emperor Bend shows change in direction of plate movement 40 million years ago. Neighboring island chains (Tuamotu line and Gilbert Austral) show similar trends. (After various authors)

Volcanic island with fringing reef   Island sinking, barrier reef   Atoll

CORAL REEFS develop around islands, especially the volcanic islands of the Pacific, where they may form fringing reefs (growing on the fringe of the island), barrier reefs (separated from the island by a lagoon), or atolls (ringlike ramparts of lagoons, with no main island core). One hundred and fifty years ago, Charles Darwin suggested these three types of reefs represented three stages in the sinking of the ocean floor around a volcanic island, the coral growth at sea level keeping pace with the sinking of the floor.

Geophysical surveys and borings on Pacific atolls confirm this theory. On Eniwetok, 4,000 feet of coral sediments overlie olivine basalt. It may not be a complete explanation, however, because of the generally uniform depth of lagoons (from 150 feet for the smaller ones to 270 feet for the larger ones), which may be the result of Pleistocene erosion during periods of lower sea level. Sea-floor spreading (p. 140) accounts for the subsidence and lateral movement of the ocean floor involved in the formation of reefs and volcanoes.

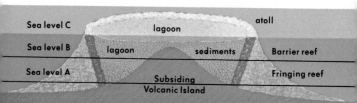

Sea level C   lagoon   atoll

Sea level B   lagoon   sediments   Barrier reef

Sea level A   Subsiding   Fringing reef
Volcanic Island

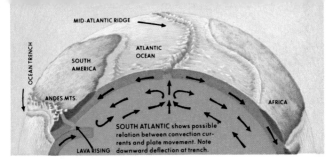

MID-ATLANTIC RIDGE

OCEAN TRENCH

ATLANTIC
OCEAN

SOUTH
AMERICA

ANDES MTS.

AFRICA

SOUTH ATLANTIC shows possible
relation between convection currents and plate movement. Note
downward deflection at trench.

LAVA RISING

PLATE TECTONICS theory offers a unified explanation for most features of the earth. The earth's surface consists of seven rigid, moving, interacting plates and several minor ones, each about 100 kilometers (60 miles) thick, carrying both continents and oceans. New crust is created at spreading ridges (*constructive* or *divergent margins,* p. 145); it moves away from these at speeds of up to 18 centimeters a year, and is destroyed by *subduction* into the mantle at *destructive* or *convergent margins,* such as trenches, where one plate typically overrides another (p. 145). Plates may also slide laterally past one another at *transform faults* (p. 141). The process seems to be continuous and broadly balanced, so the earth does not change in size.

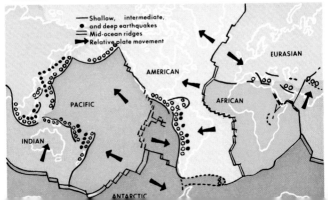

— Shallow, intermediate,
● and deep earthquakes
▥ Mid-ocean ridges
➤ Relative plate movement

EURASIAN

AMERICAN

AFRICAN

PACIFIC

INDIAN

ANTARCTIC

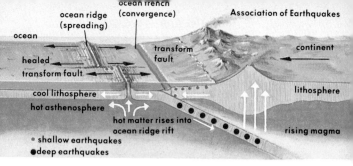

ocean ridge (spreading)

ocean trench (convergence)

Association of Earthquakes

ocean

transform fault

continent

healed transform fault

cool lithosphere

lithosphere

hot asthenosphere

hot matter rises into ocean ridge rift

rising magma

• shallow earthquakes
● deep earthquakes

OCEAN RIDGE topography, vulcanism, shallow earthquakes, and symmetrical magnetic anomalies reflect lateral spreading of new crustal material from ridge areas. (After Press and Siever)

ISLAND ARC TRENCHES reflect the subduction of oceanic plate; melting and rising material contributes to *volcanoes*; friction from downward plate movement produces *earthquakes*.

MAJOR EARTH FEATURES are explained by plate tectonics. For example, young oceanic rocks (less than 175 million years old) reflect subduction of older ones. Older continents (up to 3.8 billion years) are generally too "light" to be subducted. Other examples are illustrated here.

MOUNTAIN BUILDING takes place at *convergent boundaries* of plates, where collision produces intense compression. There are several varieties of collision.

ANDES arise from subduction of the oceanic Nazca plate below the continental South American plate. Mountain building, vulcanism, earthquakes, uplift, and the deep Chile trench result, as in figure above.

HIMALAYAS formed from collision 40-60 million years ago of Indian and Eurasian plates; overthrusting produces mountains. The Alps reflect collision of African and Eurasian plates, 60-80 million years ago.

PERMIAN GLACIAL DEPOSITS, now scattered across southern continents, were formed within a supercontinent, Gondwanaland. This drifted apart, and the individual continents are now isolated by plate movement.

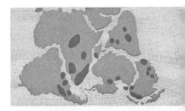

HOW THE EARTH WORKS

IF SEAFLOOR SPREADING and the interaction of the rigid plates of the earth's crust (plate tectonics) can account for all the major features of the earth, we have to ask what can account for the various phenomena of plate tectonics. What is the driving force that moves the huge, rigid slabs that make up the surface of the earth? How can any such force be powerful enough to cause earthquakes and raise up mountains? What is the earth engine that drives plate tectonics?

Most earth scientists now conclude that the mantle (p. 129)—the dense layer underlying the crust—is a solid, so hot that it can flow very slowly. "Floating" on that foundation, the lighter crustal plates are dragged along by these slow movements in the mantle.

How could that work? There is no agreement on this, but three possible ways are described on p. 147. Whatever movement is proposed must be capable of explaining *lateral* movement of plates away from spreading ridges and *downward* movement of plates at trenches.

SIMPLIFIED MODEL of how plates of crust may be driven by convection currents in the mantle. (After several authors)

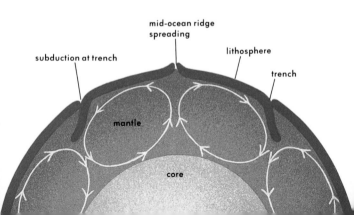

FORCES THAT MOVE PLATES are still not clear. Illustrated below are three possible models.

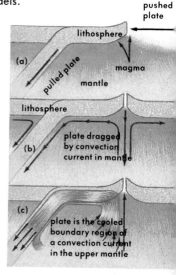

pushed plate

PUSH-PULL FORCES may produce plate movement. The height and weight of spreading ocean ridges may push the plate laterally, while the cool, heavier subducted plate may pull in the same direction.

CONVECTION CURRENTS in the mantle may form several slowly convecting cells, rising below the ridges, sinking below the trenches, dragging the crust with them.

PLATES may be cooled portion of upper mantle formed by convection currents rising at ridges and cooling as they spread. (Illustrations after Press and Siever)

EARLIEST HISTORY OF THE EARTH is still obscure, but it probably involved the accretion about 4.7 billion years ago of cold materials. The impact of this material gradually heated up the growing earth, as did compression and radioactivity. The temperature slowly increased until, perhaps 1 billion or so years after its formation, the earth was hot enough to melt the iron present, which sank towards the center, or core, creating more heat and causing differentiation of lighter material into crust and intermediate mantle. As the temperature rose, slow convection movement began to take place, and the differentiation into a layered earth continued, with lighter material forming the continents. The atmosphere and oceans probably accumulated by outgassing from within the earth. Plate tectonics is part of this later history of the earth.

THE EARTH'S HISTORY

The age of the earth has been a subject of speculation since early days of humankind, but only in the last century have attempts been made to measure it. Geologists studying earth processes are concerned with the sequence of rocks and structures in time, and thus with the history of the earth itself.

RATE OF COOLING was once thought to show the earth was only 20-40 million years old. Discovery of heat produced by radioactivity provides new data that invalidate this conclusion.

TOTAL AVERAGE THICKNESS of sediment over the earth, if regularly deposited, was thought to provide the earth's age, if divided by the average annual addition to new sediments. While this method may be acceptable for a few local deposits, there are far too many variables and unknowns (such as redeposition of sediment) to allow its use for determining the age of the earth.

SALT CONTENT OF THE OCEANS is presumed to have come from weathering of rocks and was divided by the annual increment to give an age for the oceans of about 90 million years. The same limitations apply to this factor as to sediment-thickness calculations. Both figures involve corrections that would greatly increase their value.

RADIOACTIVE DECAY provides the best present method of measuring the age of rocks. Radioactive elements undergo spontaneous breakdown by loss of alpha and beta particles into stable elements. The rate of breakdown, which can be accurately measured, is independent of any environmental conditions, such as temperature or pressure. The ratio of decayed to parent elements thus provides an indication of the age of the mineral in which it is found. Different elements have very different decay rates.

◄ Succession of sedimentary rocks in Grand Canyon, Arizona, shows their relative geologic ages, but represents only part of total geologic time (p. 151).

URANIUM (U238) is a commonly used element, going through 5 disintegrations before it becomes the stable element lead (Pb206). Different elements have different rates of disintegration. U238 has a half-life (the time taken for half its atoms to disintegrate) of 4,500 million (4.5 billion) years. Some other radioactive elements commonly used in age studies are thorium, potassium, and rubidium.

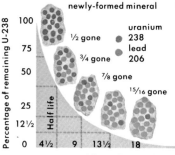

newly-formed mineral

Percentage of remaining U-238

uranium 238
lead 206

½ gone
¾ gone
⅞ gone
15/16 gone

Half life

100
75
50
25
12½
0

4½   9   13½   18

Increasing time (in billions of years)

THE OLDEST ROCKS measured by radioactive methods have an age of about 3,800 million (3.8 billion) years. This is younger than the earth itself, which is probably about 4.5-5.0 billion years old. Studies of meteorites, which are probably samples of the planetary materials from which the earth originated, all indicate an age of about 4.5 billion years. The oldest known fossils are about 3.4 billion years old, but common fossils are found only in rocks younger than 600 million years old.

Radiocarbon dating is useful for rocks that contain wood fragments and are younger than about 70,000 years. C14 from the atmosphere is incorporated in plant tissues and disintegrates to N14 with a half-life of 5,570 years.

▼

Neutrons bombard nitrogen atoms in atmosphere

cosmic rays

Rate at which C14 decays and becomes N14 is known

N^{14} N^{14} N^{14} + neutron = C^{14} (radiocarbon + proton)

0   0

$C^{14}O_2$ $C^{14} + O_2 = C^{14}O_2$ absorbed by tree

Tree   Tree absorbs $C^{14}O_2$

$C^{12}O_2$

Amounts of $C^{14}O_2$ and $C^{12}O_2$ remain constant in living tree

Section of living tree contains x amount of C^{14}

DEATH OF TREE

to atmosphere

100% C^{14}

5,700 years

50% of C^{14}

11,400 years

25% of C^{14}

17,100 years

12½% of C^{14}

UNDERSTANDING EARTH HISTORY also involves the study of the development of animals and plants and of the continuously changing ancient geographies. Particular sedimentary rocks, dated by various methods, can then be related to a general geologic time scale. Most radioactive minerals used for age determinations occur in igneous rocks, although glauconite is a mineral used for age studies of sedimentary rocks.

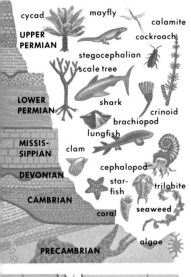

SUPERPOSITION of sedimentary rocks indicates their relative ages. In undisturbed sections, younger rocks overlie older.

STRATIGRAPHIC CORRELATION of strata in one place with those of the same age, deposited at the same period of time in another place, is fundamental in the interpretation of geologic history.

FOSSILS are important in correlation of sedimentary rocks. Fossils are the remains of, or direct indication of, prehistoric animals and plants. Although influenced by environment, similar assemblages of fossils generally indicate similarity of age in the rocks that contain them.

ROCK FACIES, the sum total of the characteristics of a rock's depositional environment, are independent of geological time. An awareness of them, however, is important in correlation. Figure shows how a shallow sea transgressed over a deltaic and near-shore environment in western U.S. in Cambrian times, about 500 million years ago.

LITHOLOGICAL CORRELATION uses the similarity of mineralogy, sorting, structure, bedding, sequence, and other features as indications of similar ages of rocks. It is of limited value, since rocks of different lithology often are deposited at the same time in adjacent areas.

GEOPHYSICAL CORRELATION makes use of similarity of physical rock properties (electrical resistivity and self-potential, for example) as an indication of similar age. Widely used in boreholes, this method is limited by the same factors as lithological methods.

STANDARD GEOLOGIC COLUMN, which has been built up by combining rock sequences from different areas, can be matched with a time scale based on measured absolute ages of rocks. This is like a master motion picture film in which local rock sequences each represent a few single frames. This geologic time scale is shown on p. 152.

ROCK SYSTEMS in the geologic column are major divisions of rocks deposited during a particular period of geologic time. Names of systems are taken from areas where rocks were first described, such as Devonian from Devonshire, or from their characteristics, as Cretaceous from the chalk, which is found in many strata of this age.

World Wars	Renais-sance		Roman Empire	Assyrian Empire	Egyptian Middle Kingdom
2000	1500	1000	A.D.	B.C. 1000	

Geologic time in millions of years (below) compared with last 4000 years

C M Pale.   Precambrian

500	1000	2000	3000	4000

Age of layers in millions of years

Sandstone
Limestone   BRYCE CANYON   60
Shale
Limestone and sandstone   ZION CANYON   135
Sandstone and shale   166
Limestone, sandstone, and shale   181

The Earth's total history is pieced together by comparison of rocks from many areas.

GRAND CANYON

230
280
300
345
600

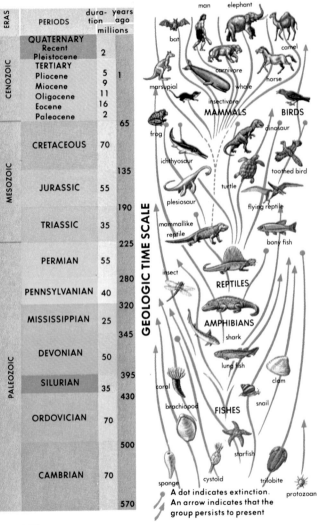

ERAS	PERIODS	duration millions	years ago millions
CENOZOIC	**QUATERNARY** Recent, Pleistocene	2	
	TERTIARY Pliocene Miocene Oligocene Eocene Paleocene	5 9 11 16 2	1
			65
MESOZOIC	**CRETACEOUS**	70	
			135
	JURASSIC	55	
			190
	TRIASSIC	35	
			225
PALEOZOIC	**PERMIAN**	55	
			280
	PENNSYLVANIAN	40	
			320
	MISSISSIPPIAN	25	
			345
	DEVONIAN	50	
			395
	SILURIAN	35	
			430
	ORDOVICIAN	70	
			500
	CAMBRIAN	70	
			570

GEOLOGIC TIME SCALE

man elephant bat camel carnivore whale horse marsupial insectivore

MAMMALS BIRDS

frog dinosaur ichthyosaur toothed bird turtle plesiosaur flying reptile mammallike reptile bony fish insect

REPTILES

AMPHIBIANS

shark lung fish clam coral snail brachiopod

FISHES

sponge cystoid starfish trilobite protozoan

A dot indicates extinction.
An arrow indicates that the
group persists to present

152

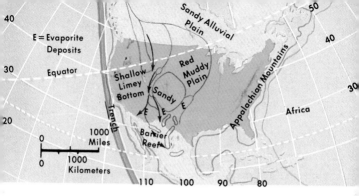

PALEOGEOGRAPHY OF U.S., in Middle Permian times, about 250 million years ago. (After Dott and Batten)

PALEOGEOGRAPHIC MAPS are reconstructions of the geography of past geologic periods. Past continental geographies can be pieced together by using paleomagnetic data and by plotting the distribution of different rock types, fossils, and geologic structures, using the methods illustrated on pp. 150-151.

WORLD PALEOGEOGRAPHY about 200 million years ago. Shading represents deposits of former ice cap (see p. 145). (After Press and Siever)

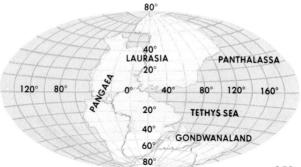

GEOLOGIC MAP OF

QUATERNARY
Rocks and unconsolidated
deposits of Pleistocene
and Recent age

TERTIARY
Rocks of Paleocene, Eocene,
Oligocene, Miocene, and
Pliocene age

MESOZOIC
Rocks of Triassic,
Jurassic, and
Cretaceous age

LATE PALEOZOIC
Rocks of Devonian,
Mississippian, Pennsylvanian,
and Permian age

EARLY PALEOZOIC
Rocks of Cambrian,
Ordovician, and
Silurian age

PRECAMBRIAN
A variety of igneous, metamorphic,
and sedimentary rocks (includes
some metamorphosed Paleozoic locally)

THE UNITED STATES

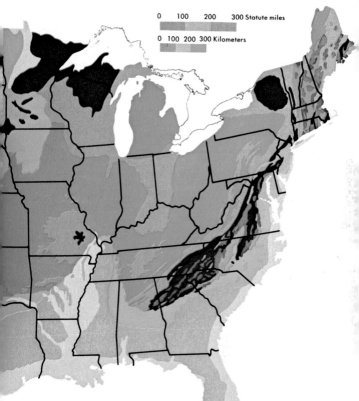

0 100 200 300 Statute miles

0 100 200 300 Kilometers

EXTRUSIVE IGNEOUS ROCKS
Chiefly lava flows of
Tertiary and Quaternary
age

INTRUSIVE IGNEOUS ROCKS
(includes some metamorphic
rocks) Granitoid rocks of
various ages

FOR MORE INFORMATION

MUSEUMS AND EXHIBITS provide excellent displays of general geologic topics and of regional geology. Many universities and most large cities have museums with geologic exhibits.

STATE GEOLOGICAL SURVEYS publish maps, guide books, and elementary introductions to geology. The U.S. Geological Survey, Washington, D.C. 20242, and the Geologic Survey of Canada also publish many useful reports and maps.

NATIONAL PARKS AND MONUMENTS are generally areas of great geologic interest. They provide guided tours, talks, museums, and pamphlets.

FURTHER READING

BOOKS will help you develop your interest in earth science. Here are a few useful titles:

Cattermole, Peter, and Patrick Moore. *The Story of the Earth.* New York: Cambridge University Press, 1985. A simply written, useful overview.

Larson, Edwin E., and Peter W. Birkeland. *Putnam's Geology,* 4th ed. New York: Oxford University Press, 1982. Well-illustrated, readable, college-level text.

Press, Frank, and Raymond Siever. *Earth,* 4th ed. New York: W. H. Freeman and Company, 1986. Comprehensive, college-level text.

Rhodes, Frank H. T., Herbert S. Zim, and Paul R. Shaffer. *Fossils.* New York: Golden Press, 1962.

Smith, David A., ed. *Cambridge Encyclopedia of Earth Sciences.* New York: Cambridge University Press, 1981.

Stanley, Steven M. *Earth and Life Through Time.* New York: W. H. Freeman and Company, 1985. College-level review of historical geology.

Zim, Herbert S., and Paul R. Shaffer. *Rocks and Minerals.* New York: Golden Press, 1957.

INDEX

158